Zeitschrift für Betriebswirtschaft

Special Issue 1/2010

Corporate Social Responsibility and Stakeholder Dynamics

ZfB-Special Issues

2/2007 Innovation, Orientation, Innovativeness and Innovation Success
Herausgeber: Holger Ernst/Hans Georg Gemünden
156 Seiten. ISBN 3 8349 0698 0

3/2007 Direct Marketing
Herausgeber: Manfred Krafft/Jürgen Gerdes
192 Seiten. ISBN 3 8349 0728 6

4/2007 Open Innovation between and within Organizations
Herausgeber: Holger Ernst/Hans Georg Gemünden
140 Seiten. ISBN 3 8349 0731 6

5/2007 Hochschulrechnung und Hochschulcontrolling
Herausgeber: Hans-Ulrich Küpper
188 Seiten. ISBN 3 8349 0778 2

6/2007 Empirische Studien zum Management in mittelständischen Unternehmen
Herausgeber: Horst Albach/Peter Letmathe
140 Seiten. ISBN 3 8349 0799 5

1/2008 Economics and Management of Education
Herausgeber: Dominique Demougin/Oliver Fabel
192 Seiten. ISBN 3 8349 0904 1

2/2008 Corporate Governance in der Praxis mittelständischer Unternehmen
Herausgeber: Horst Albach/Peter Letmathe
136 Seiten. ISBN 3 8349 0931 2

3/2008 Corporate Social Responsibility
Herausgeber: Joachim Schwalbach
112 Seiten. ISBN 3 8349 1044 9

4/2008 Forschungsperspektiven der betriebswirtschaftlichen Logistik
Herausgeber: Hartmut Stadtler/Herbert Kotzab
164 Seiten. ISBN 3 8349 1157 7

5/2008 Erich Gutenbergs Theorie der Unternehmung – Wirkungen auf die heutige Betriebswirtschaftslehre –
Herausgeber: Joachim Reese/Marion Steven
112 Seiten. ISBN 3 8349 1187 9

6/2008 50 Years after MM: Recent Developments in Corporate Finance
Herausgeber: Wolfgang Breuer/Marc Gürtler
188 Seiten. ISBN 3 8349 1426 2

1/2009 International Entrepreneurship
Herausgeber: Andreas Al-Laham/Martin K. Welge/Joachim Schwalbach
196 Seiten. ISBN 3 8349 1474 6

2/2009 Management von Familienunternehmen
Herausgeber: Peter Witt
168 Seiten. ISBN 3 8349 1620 X

3/2009 Operations Research in der Betriebswirtschaft – Neue Anwendungsgebiete und Ergebnisse
Herausgeber: Heinrich Kuhn/Hartmut Stadtler/Gerhard Wäscher
144 Seiten. ISBN 3 8349 1723 0

4/2009 Rational Inefficiencies
Herausgeber: Günter Fandel
136 Seiten. ISBN 3 8349 1856 3

5/2009 Entrepreneurial Finance
Herausgeber: Wolfgang Breuer/Malte Brettel
132 Seiten. ISBN 3 8349 2005 3

6/2009 Management von kleinen und mittleren Unternehmen
Herausgeber: Peter Letmathe/Peter Witt
180 Seiten. ISBN 3 8349 2139 4

Corporate Social Responsibility and Stakeholder Dynamics

Editor

Prof. Dr. Joachim Schwalbach

GABLER

Die Deutsche Bibliothek – CIP-Einheitsaufnahme

Zeitschrift für Betriebswirtschaft : ZfB. – Wiesbaden :
Gabler/GWV Fachverlage GmbH
Erscheint monatl. – Aufnahme nach Jg. 67, H. 3 (1997)
Reihe Ergänzungsheft: Zeitschrift für Betriebswirtschaft /
Ergänzungsheft. Fortlaufende Beil.: Betriebswirtschaftliches
Repetitorium. – Danach bis 1979: ZfB-Repetitorium
ISSN 0044-2372
2010, Special Issue 1. Corporate Social Responsibility and Stakeholder Dynamics
Editor: Joachim Schwalbach – Wiesbaden: Gabler, 2010
(Zeitschrift für Betriebswirtschaft; 2010, Special Issue 1)
ISBN 3-8349-1995-0
ISBN 978-3-8349-1995-3

Gabler/GWV Fachverlage GmbH, Wiesbaden 2010
Lektorat: Susanne Kramer/Annelie Meisenheimer

Gabler ist Teil der Fachverlagsgruppe Springer Science+Business Media.

http://www.gabler.de
http://www.zfb-online.de

Höchste inhaltliche und technische Qualität unserer Produkte ist unser Ziel. Bei der Produktion und Verbreitung unserer Bücher wollen wir die Umwelt schonen: Dieses Buch ist auf säurefreiem und chlorfrei gebleichtem Papier gedruckt. Die Einschweißfolie besteht aus Polyäthylen und damit aus organischen Grundstoffen, die weder bei der Herstellung noch bei der Verbrennung Schadstoffe freisetzen.

Satz: Fotosatz-Service Köhler GmbH, Würzburg
Printed in Europe

ISBN 3-8349-1995-0
ISBN 978-3-8349-1995-3

INHALTSVERZEICHNIS

Editorial

The papers in this special issue were presented at the 3rd International Conference on Corporate Social Responsibility at Humboldt-Universität zu Berlin, October 8-10, 2008. These papers are a fine collection of scientific contributions surrounding the topic of CSR and stakeholder dynamics. Many of the ideas in these papers are based on the incubation as early as 1984 in the book by R. Edward Freeman, *Strategic Management: A Stakeholder Approach*. But they go far beyond the basic idea that business can be understood as a set of relationships among those who have a stake in the activities that make up the business. As Freeman in his recent book about *Managing for Stakeholders* rightfully claims, business in the last two decades has witnessed unprecedented dynamics that the traditional models and frameworks cannot account. The purpose of this special issue is to present papers which intend to analyze the fast changing business environment imbedded in a very complex world with a great deal of uncertainty.

The first paper by Arora and Petrova examines the relationship between different stakeholder groups and firm performance by relying primarily on the resource dependence perspectives. The empirical results show strong validation for the resource dependence and agency theory predictions. Interestingly, they also observe strong firm performance effects by employees, consumers and diversity management but less so by proactive environment protection policies. Based on the results, Arora and Petrova suggest studying the business case for social performance at the level and dynamics of different stakeholders.

The paper by Auger, Devinney and Louviere concentrates on the impact of social product attributes on consumer purchase intentions. Many survey based studies suggest that consumers are giving increasing consideration to the social components of products and business processes and that these concerns have various implications for the businesses involved. Auger et al. claim that these surveys overstate the importance of social features due to the problem of incentive compatibility, i.e. the lack of any penalty for not revealing the truth. In other words, consumers indicate in surveys that social issues are important but do not change their purchase behaviors accordingly. Auger et al. using experimental methodologies, primarily choose experiments and best-worst scaling to deal with the problem and to gain a better understanding of consumers' 'true preferences' with respect to social and ethical issues. Their results show that most consumers do not know about the social product features of the product they purchase. Furthermore, the study demonstrates that differences among consumers are not country specific and do not relate to observable demographic factors.

In their contribution, Schneider and Schwerk focus on CSR in the supply chain of Western focal firms in China. It is commonly known that Asian suppliers often do not comply with established codes of conduct of Western focal firms. Based on semi-structured interviews, Schneider and Schwerk show that most Chinese suppliers still consider codes of conduct as external interference with their business policy and as a driver of additional costs. The interviews also reveal that dialogue-oriented approaches applied by the focal firms support the compliance with codes of conducts much more effectively than standards-based approaches.

In his paper, Scholtens explores the proliferation of fair trade in Europe. Fair trade is a global phenomenon and relates to consumers that included characteristics of sourcing and production processes into their buying decisions. As a consequence, social factors may impact investment decisions by firms. Scholtens aims to determine the drivers of fair trade at the national level and associates the size of fair trade with economic and cultural indicators. For Europe, Scholtens finds evidence that socio-cultural factors rather than economics factors appear to be associated with the development of fair trade.

The final paper by Haezendonck, Dooms and Verbeke is a case study about the dynamic process of stakeholder salience and its effect on the company DHL regarding its decision to concentrate all its daily air cargo and express flights, including night flights, at Brussels National Airport. Stakeholder dynamics is analyzed within the typology using the contracts of power, legitimacy and urgency. The case study clearly shows how the DHL's preference for expansion in Brussels was ultimately transformed into a preference for exit. The authors therefore suggest that stakeholder management analysis should move beyond finding static scores allegedly exhibited by various stakeholders for salience attributes, and take into account dynamic elements.

Prof. Dr. Joachim Schwalbach
Humboldt-Universität zu Berlin
Institute of Management

ZfB
ZEITSCHRIFT FÜR BETRIEBSWIRTSCHAFT

Corporate Social Performance, Resource Dependence and Firm Performance

Punit Arora, Milena Petrova

Abstract: Using resource dependence perspective we examine the relationship between different stakeholder groups and firm performance. Our results indicate that different stakeholder groups have different levels of salience for firm's financial performance. Proactive policies with respect to employees, consumers and diversity show the maximum beneficial effects. Our results also indicate that commitment-based work practices help firms improve their capacity to benefit from their performance on other CSP domains, which we believe indicates preliminary support for the concept of stakeholder influence capacity.

Keywords: Resource dependence perspective · CSR · CSP · relationship between CSR domains · CSR controversies

JEL Classification: M00

Punit Arora (✉)
Business Economics, Syracuse University, Syracuse NY 13244, USA;
E-mail: punit@syr.edu; URL http://whitman.syr.edu/Directory/showInfo.aspx?id=311

Assistant Prof. Milena Petrova, Ph.D. (✉)
Whitman School of Management, Syracuse University, Syracuse NY 13244, USA
E-mail: mpetrova@syr.edu; URL: http://whitman.syr.edu/Directory/showInfo.aspx?id=247

Organizations need a variety of resources for their survival and their success from a resource dependence perspective is dependent on their ability to satisfy their key stakeholders (Pfeffer and Salancik, 2003). These coalitions of stakeholders indulge in active and extensive bargaining process for exchange of resources that each party needs to best meet its goals (Cyert and March, 1963). Most resource allocation decisions within organizations are therefore an outcome of such bargaining. In this give and take, different stakeholder groups have different levels of salience and ability to achieve their goals in different contexts, which may either depend on the primacy of certain stakeholders or be subject to negotiations across coalitions of stakeholders. Most resource allocation decisions that lead to corporate social programs can be viewed in this light, yet research on corporate social performance (CSP) in the positivist domain focuses largely on establishing the link between the overarching concept of CSP and financial performance (FP), without considering the resource allocation process from the coalitional perspective (*e.g.*, Barnett, 2007).

Corporate social performance is a term that encompasses within its contours firms' various non-market activities directed at stakeholders other than its shareholders. These activities include attention to and investments on improving firm's relationship with its employees, consumers, regulators, and communities in general. Some of these activities such as building trusting relationships with its customers or implementing high-performance work practices may have a direct and immediate effect on the firm's business, others such as promoting nuclear disarmament or refusing to deal with tobacco or alcohol manufacturers[1] may have only an indirect, and in many cases a negligible, effect. Then, there is an entire spectrum of activities such as protecting environment and promoting diversity that fall somewhere in between. In general, from a resource dependence perspective, firms undertake these activities in an attempt to manage their environment: acquiring power for reducing their dependence on others or increasing others dependence on them (*e.g.*, Pfeffer and Salancik, 2003), which is critical for their survival. If the firms' goal is to prosper by maximizing the inflow and minimizing the outflow of resources, they are likely to apportion their resources in a manner that pays greater attention to stakeholders that control critical resources, and they are at the same time more likely to benefit more from it.

While over last three decades, there has been a constant flow of studies devoted to the empirical examination of the relationship between CSP and FP (*e.g.*, Moskowitz, 1972; Vance, 1975; Alexander and Buchholz, 1978; Aupperle, Carroll and Hatfield, 1985; Johnson and Greening, 1994; Waddock and Graves, 1997; Seifert, Morris and Bartkus, 2003), these studies did not pay adequate attention to coalition dynamics and stakeholder salience, and perhaps that is why their results for CSP-FP relationship have been inconsistent: strongly positive to strongly negative. Therefore, the first question we ask is: If different stakeholder groups have different salience levels for firm's performance (*e.g.*, Barnett, 2007, Wood and Jones, 1995), is it theoretically and empirically justifiable to combine firm's performance with respect to different groups of stakeholders under the rubric of a broader concept of *corporate social performance*? In other words, ***does the attention to various social performance domains have similar effect on firm's financial performance?*** In this process, we also examine the empirical validity of combining various activities in the social performance domain into a CSP index.

Second, some research (*e.g.* Zaltman and Zaltman, 2008) has suggested that employees are the critical link in the CSP-FP relationship. For example, in forming their perceptions about how much the firms value them (customers), they look at how firms treat their employees. Thus, customers tend to believe that such firms that do not treat their employees well can hardly be expected to care about them (Zaltman and Zaltman, 2008). Heskett (2008) goes even further and claims that only satisfied employees can create satisfied consumer base and lifelong profits. This is consistent with Barnett's (2007) position. He believes that employees are vital in improving the firm's relationship with all other stakeholders. His concept of stakeholder influence capacity assigns a pivotal role to the employees in developing organization's relationships with other stakeholders. In concurrence with this concept, ***we test for interactive effects between employee relations (or high powered employment practices) and other social performance domains, and assess the relationship of these interactive effects on firm's bottom line***.

1 Literature review

Over 150 studies can be cited to adduce the sharp divide in the literature on the nature of relationship CSP and FP. Here, we encapsulate only a handful of such studies. Orlitzky, *et al* (2003) conducted a meta-analysis of 52 studies from 1970 to 2003, and their conservative estimates strongly support their hypothesis that higher CSP leads to better FP. This study substantiated the findings of the classical Moskowitz (1972) study on 14 firms that found evidence of superior stock market returns for those firms that demonstrated good social behavior, and that of Waddock and Graves (1997) who found a strong CSP- FP link. Researchers like Vance (1975), on the other hand, found that the firms were punished by the stock markets for diverting shareholders' wealth to social causes. Yet, another group of researchers found no evidence for a significant relationship between CSP and FP (*e.g.*, Aupperle, *et al,* 1985; Alexander and Buchholz, 1978).

However, most of these studies were essentially carried out within narrowly defined research parameters: regression between CSP and FP with a few control variables. Most such models either miss out on certain key variables or attempt to link the concepts in a way that has no basis in theory or practice. This prompted McWilliams and Siegel (2000, p.603) to call such studies as *misspecified* models because researchers failed even to remove the effect of firms' research and development or advertising intensity, which they believe causes the results to be "upwardly biased estimates of the financial impact of CSR." Similarly, Ullmann (1985, p.540) termed such studies as "data in search of theory", Mahon and Griffin (1999, p.126) as "partial insights into an unseen larger picture", and Wood and Jones (1995, p.229) as "stakeholder mismatches" or attempts at finding "relationship between unparallel concepts".

Previous research implicitly assumed that all kinds of social performances have similar impact on the firm profitability. Thus, their research designs either use an index (*e.g.* Waddock and Graves, 1997) or a particular dimension of social performance such as corporate philanthropy (*e.g.*, Seifert, et al, 2004), and then generalize the results across board. However, these studies do not provide any theoretical justification as to why it is reasonable to coalesce together all such expenditures on issues ranging from employee incentives and consumer relations to nuclear safety and human right advancement, nor do they specify why all such expenditures would have similar effect on firm's financial performance.

It is also noteworthy that researchers, depending on their theoretical disposition, assume all social expenditures to be either good or bad. In real life, corporations commit to CSP for both *right* and *not-so-right* reasons. Sometimes managers commit to social causes to build better relationships with various stakeholders and promote the long-term sustainability of their businesses (Hillman and Keim, 2001), and other times they do it merely to further their own interests including prestige enhancement and entrenchment (Surroca and Tribo, 2006). If managers indulge in CSP for such conflicting reasons, then how do we know when shareholders would punish social performance and when they would reward it. The answer to this riddle perhaps lies in the capital markets' evaluation of CSP expenditure in relation to its impact on firm's long-term sustenance and profitability. This, of course, assumes that the capital markets are efficient enough to distinguish between good and bad social expenditures. It also assumes that they are discerning enough to price the social investments that may not exhibit direct and immediate benefits. These assumptions may appear demanding at first glance, but they form the core of the efficient markets hypothesis, which states that the markets can (at least asymptotically) correctly evaluate and price the firms' strategies and performance (Malkiel, 2003; Fama, 1998). Thus, it may be difficult for individual members to ascertain the benefits from CSP, but capital markets as a whole can and do arrive at the appropriate decisions as reflected in the stock prices, just as they do in the case of evaluating firm's other risky and uncertain projects such as entrepreneurial ventures into unproven technologies/ markets, diversification into new businesses or even acquisition of existing competitors. If this argument is valid, then the only time when the managers can get away with 'bad' or sub-optimal CSP is when the firms have weak corporate governance systems. Under conditions of good governance, the interests of the managers are well aligned with those of the shareholders. Managers get rewarded for good performance and punished and removed for bad performance. Under weak governance, however, managers are better able to exploit the situation to their benefit, which might include wasteful spending under the garb of CSP. In fact, some studies indicate that "much that is often lumped in with CSR actually falls outside its bounds (Barnett, 2007, p.799)." This implies that we need to draw a line of fine distinction between managerial acts aimed at enhancing their own prestige and the social responsibilities they are called upon to assume.

Hence, any study of the link between CSP and FP needs to take governance systems in consideration to avoid detecting spurious relationships, positive or negative. Therefore, in the present study, we consider the impact of governance systems, before studying the impact of relationship with each stakeholder group on firm profitability. Hillman and Keim (2001) initiated this process in their seminal piece on the distinction between stakeholder management and social issue participation. They found out that while the stakeholder management improved shareholders' equity, social issue participation led to its decline. In this paper, we take this concept to its next logical step, and investigate whether it is empirically valid to combine firms' performance with respect to different stakeholders under the all encompassing construct of social performance or even stakeholder management before we investigate relationship with the firms' financial performance.

2 Theory development

The resource dependence model of Pfeffer and Salancik (1978) postulates that organizations cannot internally generate all the resources they need by themselves, and therefore they must "enter into transactions and relations with elements in the environment that can supply the required resources and services" (Aldrich and Pfeffer, 1976, p.83). For ensuring a constant supply of critical resources, managers have two options. They can either gratify owners, members and powerful constituents that they are dependent on (White, 1974) or they can actively manage their external dependence. While gratifying these stakeholders involves actions such as maintaining healthy relationships, providing extensive services and/ or meeting performance expectations on key parameters, actively managing external dependencies includes, among others, using trade associations, building coalitions and co-opting key external constituents/ stakeholders. From the resource dependence perspective, the latter response is as important, if not more, as the former. Organizations are believed to be capable of changing as well responding to their environment (Aldrich and Pfeffer, 1976). In fact, not only they cater to the demands of some stakeholder groups but also try to avoid certain others. Two plausible reasons have been offered for this avoidance strategy (*e.g.*, Oliver, 1991). One, organizations attempt to negate the power and control possibilities inherent in resource dependence. Two, different stakeholder groups, including managers and shareholders, scramble to claim the largest share possible from the organizational kitty for themselves, resulting in active political hustle and bustle to include (and exclude) different claimants to the extent necessary (and possible). In summation, unlike stakeholder theory as applied in the CSP literature, which implicitly assumes that the attention to all social domains is equally salient to firm's financial performance, (*e.g.*, Waddock & Graves, 1997), resource dependence propositions that different stakeholders have different salience for firm performance because of the difference in resources that they can offer and contribute and the differences in bargaining power that they command. Therefore, to obtain more consistent results, theoretically[2] it appears more justifiable to study the CSP-FP relationship at the level of different stakeholder groups.

Next, we develop hypotheses in tune with this line of inquiry, and examine the relationship between six domains of CSP that previous research has studied as CSP index (*e.g.*, Waddock & Graves, 1997) and financial performance. These components are employee relations, consumer relations, environmental protection, diversity promotion, community relations and other social performance domains. In deriving these hypotheses, we extensively draw upon relevant literatures from their respective fields.

2.1 Employee relations and performance

Effective employee relations involve instituting commitment based work practices such as extensive employee involvement in decision-making, training and professional development, appropriate compensation and incentives, profit sharing, fair treatment of employees (including their unions if applicable), and long-term commitment to employees at the workplace. These work practices have long been recognized to significantly influence corporate financial performance (*e.g.*, Barney and Wright, 1998; Delaney and Huselid, 1996), and in competitive and increasingly knowledge-based economies, their importance is only going to increase! Employees control a resource critical to the firm is performance: their knowledge and skills, they also work in close geographical or psychological proximity, have access to decision-makers, and thus have the capacity to not only significantly influence firm's short-term performance but also its sustainable competitive advantage.

Therefore, as expected, this concept receives a strong endorsement from the resource-based view (RBV), the theoretical framework most commonly used to study this relationship. RBV propositions that the firms need to develop bundles of unique, valuable, and inimitable resources for developing and sustaining competitive advantage (Barney, 1991); effective HRM practices enable the firms in achieving this task rather well by fostering an environment that encourages higher organizational commitment and lesser replacement costs of tacit knowledge, skills and abilities that employees develop on the job (*e.g.*, Guthrie, 2001). These practices mould the organizational climate by shaping employees attitudes, perceptions and behaviors. Some researchers offer even more seminal reasons and attribute this to causal ambiguities and social complexities. From this perspective, it is the complexity and intricacy associated with HRM practices that makes it difficult for the competitors to imitate them (*e.g.,* Wright, McMahan and McWilliams, 1994). One of the major sources for this complexity is that HRM practices such as, decentralization, delegation of authority, and team building, help better discovery and utilization of local knowledge, which in turn may result in "non-trivial improvements" (Laursen and Foss, 2003, p.248). Thus, efficient HRM systems have the profound effect of converting firms' employees into strategic assets for a sustainable competitive advantage (*e.g.*, Amit and Shoemaker, 1993).

A plethora of empirical evidence corroborates that the use of commitment-based HRM practices can help create sustained competitive advantage, especially when they are aligned with the firm's competitive strategy, they can reduce employee turnover and increase employee motivation level, reduce waste and increase productivity, and thus improve financial returns (*e.g.*, Guthrie, 2001; Barney and Wright, 1998). In fact, Huselid (1995) found a one-standard-deviation increase in high performance work practices results in an increase of over $27,000 in sales and over $3,800 in profits! The empirical evidence from studies replicated abroad has been equally supportive (*e.g.*, Fey, Bjorkman and Pavlovskaya, 2000; Guthrie, 2001). Given such preponderance of evidence, we expect that attention to and investment in employee relations will significantly improve firm performance. Thus,

Hypothesis 1: Commitment-based relations with the employees lead to higher financial performance.

2.2 Consumer relations

Consumption is the primary means and reason for the interaction between an enterprise and society. A business firm would have no reason to exist if it did not have something to offer to the consumers. Since consumption involves an economic exchange between the firm and its customers that is necessary to support all other activities of the firm, good relations with the consumers are quintessential for the survival of the business. Moreover, satisfied customers not just support the firm, but also recommend it to other potential consumers.

The marketing literature generally relies on the resource dependence perspective (known as principle of reciprocity) to investigate the relationship between consumer relations (or customer satisfaction) and firm performance. Reciprocity implies an investment of time, effort and other irrecoverable resources with the aim of creating psychological ties that motivate the parties to maintain and set expectations for the relationship (Wulf, Odekerken-Schröder and Iacobucci, 2001, 2001). When applied to consumer-marketer relationships, it implies that the latter makes a sincere effort on providing high quality products and services. It also implies that marketers invest time and effort in ensuring the welfare of the consumers, focusing on product safety, ease of use, after-sale services etc. Bennett (1996) argues that the consumers' commitment to a marketer depends on their perception of efforts made by that marketer; the extent to which consumers want to maintain this relationship is reflected in the brand loyalty that they display.

Marketers have been quick to notice this trend towards more relationship-based exchanges with the customers; similarly researchers have acknowledged the phenomenal growth of relationship marketing (Gundlach and Murphy, 1993). Some commentators (*e.g.*, Dwyer, Schurr and Oh, 1987, p.14) call this the trend towards "the marriage of the buyer and seller." This has, at least to some extent, been in response to the realization that economic transactions are embedded in the social relationships, and that they are governed not just by price but by a combination of price, authority and trust (*e.g.*, Granovetter, 1985). Therefore, it is not surprising to notice marketers paying more attention to the relationship aspects of marketing.

Researchers conclude that consumers feel obligated to repay marketers for their 'friendliness', and empirical evidence on hand substantiates that firms with satisfied customers enjoy greater customer loyalty, positive word of mouth, reduced cost of returns and rework, lesser complaints and reduced future transaction costs, premium prices, higher levels of cash flows and reduced price volatility, and thus superior overall market value (*e.g.*, Anderson and Sullivan, 1993; Luo and Bhattacharya, 2006; Fornell *et al*, 2006). Hence, in view of the preponderant empirical evidence, we hypothesize that:

Hypothesis 2: Better consumer relations lead to superior financial performance.

2.3 Environmental protection

Environment protection includes activities such as pollution prevention programs, recycling programs, clean energy usage, innovative remediation product features, hazardous waste and emissions management programs, and environment-friendly maintenance of property, plant and equipment. In an institutional environment, where firms are under strong pressure to adopt proactive environment policies to minimize their carbon footprint, prevent damage to environment and be more concerned about climate change, resource dependence perspective propositions that "organizations can enhance their legitimacy and social fit" by responding to these salient social concerns and normative pressures (Ingram and Simons, 1995, p.1468). Even the resource-based view anticipates that those firms that voluntarily meet and exceed social concerns for the climate change would be in a better position to develop sustainable competitive advantages. From these perspectives, we would expect a positive relationship between proactive environmental policies and firm's financial performance.

However, previous research also attaches some important riders to this proposition. First, not all firms are subjected to these institutional pressures to comply with such proactive policies. The large and visible ones receive greater attention, while others escape scrutiny to a great extent (Amato and Amato, 2007). However, large firms also possess greater resources to counter institutional pressures. If they can successfully manage their environment, they can manage with the least acceptable performance levels. In competitive conditions, this signifies that the firms that are more successful in passing the *cost of the commons* on to the society, without attracting negative attention, may succeed in outperforming, at least in the short run, those that voluntarily or involuntarily choose to be proactive. Second, as we discuss later, not all firms have the capacity to gain from proactive environmental performance. In the absence of well-developed stakeholder influence capacity, firms may fail to make a persuasive case to their customers, owners and other key stakeholders (Barnett, 2007). Thus, for example, consumers may look at firms' environmental performance as a mere marketing gimmick or owners may consider it unnecessary for shareholders' wealth maximization. If that occurs, firms may fail to benefit from their proactive environmental performance. In a similar vein, Reed (2001, p.4) states that it is not a "generic argument that corporate sustainability strategies are the right choice for all companies in all situations, but rather something that must be carefully honed to the specific circumstances of individual companies operating in unique positions within distinct industries."

The empirical evidence on proactive environmental protection and firm profitability is rather scarce. While Russo and Fouts (1997) found a positive relationship between environmental performance and profitability, Halme and Niskanen (2001) found that the immediate stock market reaction to announcements of environmental investments was negative, however prices tended to return to normal levels after a certain level of goodwill was achieved. King and Lenox (2002, p.289) found strong evidence that waste prevention leads to financial gain, but they also found no evidence for increased firm profitability from "reducing pollution by other means". Despite the limited empirical evidence, we assume that given the social salience of environmental issues, proactive environmental policies lead to superior firm performance. Thus, we hypothesize as follows:

Hypothesis 3: Proactive environmental policies lead to superior profitability.

2.4 Diversity promotion and management

Firm's diversity promotion and management practices include activities such as workforce recruitment, development and promotion policies, protecting gender, disability and gay rights, creating minority friendly environs, provision for childcare, and promoting balanced work-life situations. Theoretically, the business case for proactive diversity promotion and management in workforce is similar to the argument for adopting commitment-based work practices. Diverse members bring heterogeneous experiences and opinions to the discussion, stimulate creativity, and thus contribute to better quality decision-making, especially in novel situations (*e.g.*, Richard, 2000). Diversity also allows the firms to match the demographic profile of their customers, and thus gain a competitive edge over their rivals. Empirical studies such as by Richard, Murthy and Ismail (2007) and Richard (2000) found significant empirical support for the positive relationship between proactive diversity promotion and management and firm profitability, especially over the long run. In view of this evidence, we hypothesize that:

Hypothesis 4: Proactive diversity policies lead to superior profitability.

2.5 Community and other stakeholder relations

There are a number of other *social* performance activities such as making philanthropic contributions, supporting human right issues, promoting welfare of indigenous people, avoiding South Africa during apartheid era, refraining from economic activities in problem areas (say buying diamonds from conflict zones), and staying away from arms, gambling or liquor suppliers that firms undertake or are called upon to undertake. These activities are what Hillman and Keim (2001, p.126) term as "social issue participation". While often laudable in nature, the theoretical argument and empirical evidence as to why these activities will lead to superior firm financial performance is on weak ground for two reasons.

First, from a resource dependence perspective, these stakeholders do not have much to contribute to firm's performance. They rarely control any critical resources or possess strong bargaining power. Therefore, it is easier for the firms to avoid higher performance on this social domain without attracting any significant penalty. In fact, if managers take too much interest in such activities, they might attract negative response from owners (for reduced profits) and consumers (for higher product prices). Owners and institutional investors are especially likely to express their resentment for such activities and penalize the firm in the capital markets.

Second, while for the majority of firms their consumers are geographically well-dispersed, most of the activities, outlined above, tend to be locally concentrated. For example the multinational firm Proctor and Gamble sells worldwide, but much of its philanthropic activity is concentrated in Cincinnati. Similarly, Carrier's philanthropy is concentrated in Central New York. A prominent overseas example is Tata Steel whose philanthropy is concentrated in Jamshedpur, India. These examples can be multiplied, but the key idea is that firms are under tremendous pressure to cater to the demands of the communities they are located in. Firms' executives interact with the local leaders on a daily basis, and it is hard to turn down too many demands. Yet, these activities do not directly touch the lives of much the firms' consumer base, and therefore cannot be expected to yield direct and significant payoffs.

Third, the ability of the firm to benefit from its social performance may depend on the congruence between the values held by its customer base and the initiative supported by the firm. Thus, for example, in a study it was found that under certain circumstances, CSP may actually decrease consumers' intentions to buy companys' products, for example, if it impacted on the firms' ability to provide good quality products at reasonable prices (Sen and Bhattacharya, 2001). This study found that the CSP issues that a company chooses to focus on, quality of its products, consumers' personal support for those CSP issues and their general beliefs about CSP functioned as key moderators of consumers' responses to CSP. These results highlight that under certain conditions there actually might be a trade-off for the firms in choosing which of their stakeholders, consumers or underprivileged communities, they want to satisfy. Empirical evidence suggests that such social issue participation may not help the firm improve its financial performance (Hillman and Keim, 2001). In tune with this evidence, we hypothesize that:

Hypothesis 5: Better performance on community relations and other social performance domains is not related to improved financial performance.

2.6 Interaction between employee relations and other CSP domains

As stated before, employees are perhaps the most critical resource for the success of the firm, and commitment based employment practices promote superior financial performance. In addition to their direct benefit, such employment practices also improve the expected benefits from other social performance domains. Delery and Doty (1996) believe that because behavior is a function of ability and motivation, organizations that set up such HR practices ensure that individuals with the required abilities are hired and retained, and that apart from building skills, in agency theory terms, instituting commitment-based employee relations helps the firm align the interests of the employer (principal) and the employees (agents), which implies that firms seeking to benefit from social performance should focus on developing their human resources, especially with respect to their stakeholder management skills. Thus, for example, Zaltman and Zaltman (2008) underline the importance of treating employees well for developing good consumer relations. This is because, in their view, consumers in forming their perceptions about how much firms value them/take cue from how firms treat their employees. Furthermore, consumers do not trust firms that do not treat their employees well because they feel that if the firms do not care for their employees, they can hardly be expected to care for them.

Besides, commitment based work practices also develop employees as valuable and inimitable resources that can provide sustainable competitive advantage to them. Such high quality human resources would likely, directly or indirectly, help the firm in developing stakeholder influence capacity that Barnett (2007) feels is necessary for benefitting from social performance efforts. "Just as the ability of a firm to notice, assimilate, and exploit new knowledge depends on its prior stock of knowledge, firm's ability to notice and profitably exploit opportunities through CSP depends on its prior stakeholder relationships (Barnett, 2007, p.803)." While trust and reputation is important in building stakeholder relationships, it cannot be bought over the counter. It rather needs to be built over a period of time. Satisfied employees are for effective dispensers of firm's reputation, and thus contribute to the organization's ability to develop meaningful relationships with other stakeholders. Thus, firms look to keep their employees satisfied and in turn expect satisfied employees to develop relationships with other stakeholders.

Ours is a first study to investigate the interactive effects between employment practices and other social performance domains. Even though we do not directly measure stakeholder influence capacity, studying these interactive effects provides first cut at indirect empirical test of stakeholder influence capacity in the sense that we look at the impact of instituting commitment based employment practices simultaneously with paying attention to other social performance domains. Consistent with our previous theorization, we expect interactive effects to be present only in respect of developing relationships with salient stakeholders. Thus, we hypothesize that:

Hypothesis 6: Interactive effects between commitment based work practices and salient social domains (consumer relations, environment and diversity) lead to improved financial performance.

Hypothesis 7: Interactive effects between commitment based work practices and non-salient social domains (community relations and other stakeholder domains) are not related to improved financial performance.

3 Data and measurement

3.1 Research design

We draw our sample from the S&P 500 and KLD Domini 400 Universe. First, we collected data for all the 695 firms that have been continuously rated on social performance for the years 2001 to 2005 by the firm Kinder, Lyndenberg and Domini (KLD). Next, we obtained information on institutional ownership, corporate governance and financial performance for these firms. We obtained information on institutional ownership from the CDA/ Spectrum Thomson Financial's 13F database, corporate governance from the Risk-Metrics database, and financial performance from the Compustat North America database. We lagged the independent variables by a year to ensure that the independent variables predated the dependent variables. Thus, the information on the dependent variables was collected for the years 2002 to 2006. Missing or incomplete information brought our sample down to 2090 observations for 418 firms.

All the variables were standardized[3] before conducting the regression analysis. We employ time-series, cross-sectional regression analysis (random effects model), because OLS cross-sectional studies can lead to biased or misleading estimates (George, 2005; Finkelstein and Boyd, 1998). The use of a longitudinal methodology enables us to isolate the effects of specific actions and treatments over time and across sections (Hill and Phan, 1991). Moreover, the use of a random-effects model is preferable from the perspective of generalizing the findings beyond the sample under study (Maddala, 2002).

3.2 Dependent variable

We use two measures to capture firm's financial performance: one accounting and one market-based financial measure. Return on Assets (ROA) was used as an accounting measure of performance, which is a fairly standard measure in past research (e.g. Waddock and Graves, 1997). We used Market to Book Ratio (MBR) as a measure for market-based performance measure. MBR is a commonly used, modified version of Tobin's Q (*e.g.*, Richard, *et al* 2007), which is used because it is difficult to assess the replacement cost of assets, the denominator in the Tobin's q equation. MBR was chosen because it is a measure that captures the relative success of firms in maximizing shareholder value through efficient allocation and management of scarce resources[4].

3.3 Independent variables

We use the archival ratings on Corporate Social Performance (CSP) domains as our independent variables. This information was obtained from the firm Kinder, Lyndenberg and Domini. The use of KLD ratings on corporate social performance is fairly common and standard in the literature (*e.g.* Waddock and Graves, 1997). KLD provided ratings on all the firms belonging to Standard and Poor 500 or KLD Domini 400 universe for the period of 2001-05. We used ratings on community relations, employee relations, diversity, environment, consumer relations, human rights and other concerns from this database. However, unlike Waddock and Graves (1997), we did not combine these ratings into a single index.

3.4 Control variables

Previous research has indicated the need to control for industry, firm size, R&D intensity and corporate governance structures (Ullmann, 1985; Waddock and Graves, 1997; McWilliams and Siegel, 2000). Industrial munificence affects firm profitability, which in turn may affect CSP. Further, there are significant differences between different industries on the spending on research and development and expectations for social investments.

3.4.1 Industry

To control for difference is industrial munificence, we followed the 13 industries classification of Waddock and Graves (1997). We have a fairly representative cross-section of the American economy in our sample. About 19.6% of the firms belong to computers, autos and aerospace industry, 11.5% each are from telephone and utilities and bank and financial services: the rest are spread across the spectrum of remaining industries quite evenly.

3.4.2 Firm Size

Firm size is important since larger firms tend to attract greater attention and pressure to respond to the stakeholders' demands (Amato and Amato, 2007; Burke, *et al.*, 1986). While most of the previous research has measured size as total assets of the corporation, we prefer to measure size by the number of employees working for the firm. We choose this measure because we use ROA and R&D intensity, which involve total assets and sales in their denominator, which can potentially cause multicollinearity. The use of number of employees averts this problem without any loss of information.

3.4.3 Risk tolerance

Long-term debt to equity ratio is used to measure firm's risk tolerance. This again is a fairly standard and extensively used proxy for risk tolerance capacity of the firm.

3.4.4 *Research and advertisement consciousness*

McWilliams and Siegel (2000) have shown research and development intensity and advertising intensity to be other important variables that have a bearing on the relationship between CSP and CFP. They believe that a model of relationship between CSP and CFP that does not include these two variables is misspecified and fundamentally flawed since such a relation could simply reflect the impact of R&D or advertisement on firm performance. Therefore, we also control for R&D intensity, which is defined as R&D expenditure divided by sales. We also control for marketing intensity to account for firm's efforts in brand promotion, which is defined as Selling, General and Administrative Expenses divided by sales. We used this broader measure to avoid unnecessary loss of information due to paucity of information on advertising expenditure in the Compustat database.

3.4.5 *Corporate governance*

Corporate governance, broadly considered, refers to formal structures, informal structures, and processes that exist in oversight roles and responsibilities related to the ratification and monitoring of managerial resource allocation decisions (Hambrick, Werder and Zajac, 2008). It is an important facet of corporate life, and lies at the heart of investment decisions. In a 2002 survey of over 200 institutional investors that collectively managed over 9 trillion dollars in investments, McKinsey & company found that investors, in making their investment decisions, placed corporate governance at least on par with the indicators of financial performance. Further, more than 60% of them stated that they completely avoid investing in the companies with poor governance. The survey also found that an overwhelming majority (nearly 80%) of investors were willing to pay a governance premium of 12 to 14% in the matured markets of North America and Western Europe and up to 30% in East Europe and Africa (McKinsey & company, 2002). The importance attached to the governance structures is not surprising. Apart from lower risks for the investments, studies such as by Gompers, Ishii and Metrick (2003) have shown that investors earn higher returns from the investment in the better-governed companies. In fact, Gompers, et al (2003) showed that investors make as much as 8.5% in abnormal returns for reduction in the agency costs. Less dispersed stockholdings, better supervision, mechanisms to throw out the ineffective or self-serving management and such other measures of effective corporate governance have been shown to reestablish the link between the ownership and control severed by the growth of large corporations, also known as the managerial revolution (e.g. Davis, 1991). Thus, strengthening or weakening of the agency control can have a decisive influence on firm's performance, and we need to preclude the influence of corporate governance on the relationship between CSP domains and firm performance. From this perspective, we expect that under weak governance systems, firms will fail not only to squander resources on questionable social performance programs, but will also fail to fully benefit from good social performance programs. However, firms with strong governance systems will judiciously invest in social performance and harness such investments to the maximum potential.

To code for corporate governance in different firms, we use three approaches. First, concentrated ownership may be the most effective mechanism to reduce agency problems

Tab. 1. Descriptive Statistics

	Mean	S.D.	ROA	MBR	Size	DE	R&D	SGA	IO
Return on Assets (ROA)	0.04	0.12	1.00						
Market to Book Ratio (MBR)	3.58	21.12	0.09 ***	1.00					
Firm size (size)	16.57	1.39	0.06 *	0.00	1.00				
Debt-Equity ratio (DE)	0.89	11.27	0.01	0.00	0.02	1.00			
SGA intensity (SGA)	0.04	0.10	-0.23 ***	0.00	-0.22 ***	0.00	1.00		
R & D intensity (R&D)	0.20	0.19	-0.09 ***	0.03	-0.16 ***	-0.02	0.67 ***	1.00	
Institutional Ownership (IO)	0.69	0.15	0.12 ***	0.03	-0.02	0.00	-0.01	0.00	1.00
Governance index (GIM)	9.86	2.51	0.06 **	0.00	0.01	0.04	-0.14 ***	-0.10 ***	0.02
Proportion of outside directors (OD)	26.60	34.95	0.02	-0.01	0.00	-0.01	-0.08 ***	-0.10 ***	-0.06 **
Employee Relations (ER)	-0.09	16.75	0.10 ***	-0.02	-0.07 ***	0.04	0.12 ***	0.10 ***	-0.09 ***
Consumer Relations (CR)	-5.43	13.48	0.03	-0.04	-0.33 ***	-0.01	0.05 *	0.09 ***	0.10 ***
Environment protection (ENV)	-2.57	11.87	0.02	0.01	-0.12 ***	-0.03	0.09 ***	0.22 ***	0.13 ***
Diversity promotion (DIV)	11.81	16.29	0.09 ***	0.00	0.30 ***	-0.01	0.01	0.10 ***	-0.05 *
Community Relations (CMR)	2.97	10.08	0.05	0.01	0.21	-0.02	0.01	0.11	-0.07
Other CSP (OSP)	-7.96	9.78	0.01	-0.01	-0.26 ***	-0.01	-0.10 ***	-0.03	-0.06 ***

	GIM	OD	ER	CR	ENV	DIV	CMR
Governance index (GIM)	1.00						
Proportion of outside directors (OD)	0.06 **	1.00					
Employee Relations (ER)	-0.09 ***	0.04	1.00				
Consumer Relations (CR)	0.00	-0.09 ***	0.08 ***	1.00			
Environment protection (ENV)	0.03	-0.01	-0.01	0.17 ***	1.00		
Diversity promotion (DIV)	-0.04 *	-0.02	0.11 ***	-0.18 ***	0.03	1.00	
Community Relations (CMR)	-0.02	-0.02	0.07	-0.12	0.11 ***	0.34 ***	1.00
Other CSP (OSP)	0.04 †	0.00	0.02	0.18 ***	0.11 ***	-0.13 ***	-0.07 ***

Significance levels: † p <.10 * p <.05 ** P <.01 *** p< .001

Tab. 2. Regression Results

Random-effects GLS regression	Return on Assets (ROA) Coefficients with significance			Market to Book Ratio (MBR) Coefficients with significance		
	Model 1	**Model 2**	**Model 3**	**Model 4**	**Model 5**	**Model 6**
Constant	0.12	0.02	0.04 *	0.11	-0.14	-0.07 *
Industry dummies						
Firm size (size)	-0.08 †	-0.14 ***	-0.13 **	0.02	0.03	0.03
Debt-Equity ratio (DE)	0.01	0.01	0.01	0.27 ***	0.27 ***	0.27 ***
SGA intensity (SGA)	0.07	0.02	0.02	0.17 ***	0.15 ***	0.15 ***
R & D intensity (R&D)	-0.17 ***	-0.17 ***	-0.17 ***	-0.08 **	-0.08 **	-0.08 **
Institutional ownership (IO)	0.13 ***	0.13 ***	0.13 ***	0.02	0.02	0.02
Outside director proportion (OD)	0.07 **	0.05 **	0.05 **	0.04	0.04	0.04
Governance index (GIM)	0.01	0.02	0.02	-0.04	-0.04	-0.04
Employee Relations (ER)		0.09 ***	0.08 **		0.02	0.03 †
Consumer Relations (CR)		0.02	0.02		0.04 †	0.04 †
Environment protection (ENV)		-0.01	0.03		-0.03	-0.03
Diversity promotion (DIV)		0.12 ***	0.11 ***		0.04 †	0.03 †
Community Relations (CMR)		0.03	0.04		0.04	0.04
Other social performance (OSP)		-0.04 †	-0.04		-0.01	-0.02
ER x CR			0.03 †			0.04 †
ER x ENV			0.04 **			0.03 *
ER x DIV			0.02			-0.03
ER x CMR			-0.01			0.01
ER x OSP			-0.02			0.00
R^2	0.07 ***	0.11 ***	0.14 ***	0. 09 ***	0.14	0.15
Change in R^2		0.04 ***	0.03 ***		0.05 ***	0.01 **

Significance levels: † $p < 0.10$ * $p < 0.05$ ** $P < 0.01$ *** $p < 0.001$

(*e.g.*, Morck *et al*, 1988; Hoskisson *et al*, 2002), and financial institutions have both better incentives and better means to monitor mangers. Therefore, we use the proportion of total institutional ownership as a governance variable in our model. Second, greater outside director representation provides an important characteristic of a corporation's interest in both improving governance and seeking external legitimacy (Ahmed and Duellman, 2007). Therefore, we include the proportion of outside directors as a corporate governance variable in our model. Third, we borrow Gompers et al.'s (2003) governance index, which measures takeover defenses at different corporations and is fairly common in research on corporate governance (e.g., Bebchuk and Cohen, 2005). The data on institutional ownership was collected from Thomson Reuters' 13f database and information on other corporate governance variables was obtained from RiskMetrics (IRCC) database.

4 Results

We report the means, standard deviations, and correlations in Table 1. The mean value for ROA is close to 4% with a significant standard deviation of 12%. In addition, ROA relates positively to MBR, firm size, R&D intensity, institutional ownership, employee relations and diversity promotion. The descriptive statistics, including the correlation matrix, with respect to the other measures also appear in the table.

In Table 2, we provide the results of the random-effects regression model on standardized (normalized) variables for the period 2001–2005.[5] Models 1-3 show the effect of performance on different social domains on ROA and Models 4-6 on MBR. Models 1 and 4 show the results for the regression with control variables alone, whereas Models 2 and 5 include the results for social performance domains, and Models 4 and 6 represent the complete regression models, including the interaction effects.

In hypothesis 1, we predict that commitment-based relationships with the employees lead to superior financial performance. Both Model 3 ($\beta = 0.08$, $p < .001$) and Model 6 ($\beta = 0.03$, $p < .10$) support our hypothesis. Thus, we conclude that commitment-based employee relations help the firms improve their financial performance. Hypothesis 2 predicts that better consumer relations lead to superior FP. Model 6 ($\beta = 0.07$, $p < .05$) supports our hypothesis, but the coefficient is not statistically significant in Model 3. Thus, we tentatively conclude that better consumer relations help firms improve their FP. In hypothesis 3, we visualized that proactive environmental policies lead to better FP. Both Model 3 and Model 6 report nonsignificant results, thus we fail to conclude about the nature of relationship between proactive environmental policies and firm performance. Hypothesis 4 predicts that proactive diversity management leads to superior financial performance. Model 3 ($\beta = 0.11$, $p < .001$) provides evidence for a strong positive relationship with ROA, while Model 6 ($\beta = 0.03$, $p < .10$) reports a marginal support for the relationship. Thus, we find substantial support for hypothesis 4.

According to hypothesis 5, performance on community relations or other CSP domains is not related to superior financial performance. Both Model 3 ($\beta = 0.04$, $p > .10$) and Model 6 ($\beta = 0.04$, $p > .10$) report nonsignificant results for community relations. However, contrary to our expectations, Model 3 ($\beta = -0.04$, $p < .10$) and Model 6 ($\beta = -0.02$, $p > .10$) find a marginal but negative relationship between other CSP domains and FP. While nonfindings cannot be inferred as support for hypothesis, we believe it does indicate the likely nature of relationship between two constructs.

Hypothesis 6 predicts that the interactive effects between commitment based work practices and salient social domains (consumer relations, environment and diversity) lead to improved financial performance. The interactions between commitment-based employee relations and consumer relations in Model 3 ($\beta = 0.03$, $p < .10$) and Model 6 ($\beta = 0.04$, $p < .10$) show a positive relationship between with FP. Similarly, the interactions between commitment-based employee relations and environment in Model 3 ($\beta = 0.04$, $p < .01$) and Model 6 ($\beta = 0.03$, $p < .05$) show a positive relationship with FP. Other interactions are statistically nonsignificant. Because 2 out of 3 interactions show evidence for significant effects, we conclude a tentative support for hypothesis 6.

Hypothesis 7 predicts that the interactive effects between commitment based work practices and non-salient social domains (community relations and other stakeholder domains) are not related to improved financial performance. Interaction effects between commitment-based employee relations and community relations in Model 3 show a nonsignificant relationship with FP. Similarly, interaction effects between commitment-based employee relations and diversity management policies show a nonsignificant relationship with FP. Because both the relationships yield nonsignificant relationships, we can draw tentative conclusion regarding empirical support for the hypothesis 7.

5 Discussion, limitations, and future research directions

Our results provide a strong validation for resource dependence and agency theory predictions. We observe different levels of salience for performance on different social performance domains. Employees, consumers and diversity management predictably show the strongest effect on firm profitability. Surprisingly, effect of proactive environment protection did not emerge as strong as we had predicted. Other social performance domains, as predicted, show no effect on firm profitability. Overall, these hypotheses depict the importance of studying the business case for social performance at the level of different stakeholders. (Our factor and alpha analyses corroborate this proposition. Please refer to Annexure 1 and Footnote 1 for more details). We contend that combining various social performance domains, without much theoretical or empirical validity, could likely be a major cause for confounding the business case of social performance. Future research may either like to improve the theoretical conception and empirical methodology of social performance index or better still avoid it all together.

We find evidence for the interaction effects between commitment-based employee relations (also called high-powered or high involvement employment practices) and other salient social performance domains. Thus, though we do not measure Barnett's (2007) concept of stakeholder influence capacity, our results provide preliminary, indirect evidence that this may be a fruitful line inquiry for future research. If firms are rational actors, as the instrumental positivist perspective suggests, they likely are aware of both their business strategies and stakeholder influence capacities. Thus, beyond financial performance, they may take into account their stakeholder influence capacities before making CSP investments. We lack access to any such information for this study and provide only indirect evidence, but additional research might consider the use of direct measures to capture stakeholder influence capacity for making a better assessment of the economic leverage that a firm can extract from its social performance.

Although the use of multiple theoretical perspectives and longitudinal data are strengths of this study, it is not without limitations. First, a major limitation of this work, like the rest of CSP literature, is that our results are based on historical data. Even though our data is as recent as 2001 to 2005 and should not therefore pose a major concern, nevertheless our results are based on the prevailing relationships during the period studied. If as resource dependence perspective visualizes and as we hypothesize the effect of different stakeholder relationships on firm performance depends on the existing stakeholder salience, the relationships would change if these salience levels undergo a change. This might especially be relevant to the practice of proactive environment strategies. Thus, while our results found nonsignificant effects on firm performance, they do not imply that firms do not need to be proactive in this respect. An environmental issues gain more and more prominence and society and consumers become more and more demanding of business's compliance on key environmental issues, their importance for firm performance could quickly ratchet up. It is needless to add that it might be a while before firms reach this stage; in the meantime, they may choose to be proactive and build reputational effects or follow any of the other strategies available to them for managing institutional pressures[6].

Second, our work is based on empirical examination of macro effects. Firms pursue different strategies (for example, cost leader vs. differentiator), have different stakeholder influence capacity and operate in different political and socio-cultural environments. Therefore, our results should not be taken to imply that all firms are likely to obtain similar results by following the general trends noticed in this empirical work. Furthermore, firms are not the same when executing, supporting, and exploiting initiatives in the marketplace (Sen and Bhattaharya, 2001) and therefore should expect different results from following different strategies. In this context, it would also be important to mention that future research may consider including some proxy for firm's strategy in their empirical models to control for firm's intent. Thus, for example, some firms may just be happy selling cheaply, while others actively seek to build competitive advantage on the basis of socially responsible behavior. The results of stakeholder relationships on firm performance in these two cases are likely to be significantly different.

Third, future research may also study the preferences of firms' consumer base. Research in the marketing literature has indicated that the success of relationship marketing (consumer relations in our nomenclature) depends not only on firm strategy or its execution but also on the preferences of the individual customers (*e.g.,* Wulf, *et al.*, 2001). Thus, investing in relationships with buyers may not always be a preferable strategy. Solomon *et al* (1985) further clarify that only involved buyers are more likely to respond to relationship marketing efforts and that in low involvement situations it would probably not pay off. Thus, it might be useful to understand the customers' CSP preferences before reaching conclusions regarding their effects on firm performance.

Fourth, most of the work in the social performance literature has been carried out with multi-industry samples; future research may examine CSP with industry specific models. Such research can potentially help us understand the differences in types of social performance being carried out across different industries, which may provide some fresh insights into the CSP-FP relationships.

Lastly, there have not been any studies that have explored the CSP from an institutional theory perspective. Do mimetic and normative pressures encourage firms to follow other firms in their industry in allocating higher expenditures on CSP? How effective are the governmental regulations or coercive pressures in pushing firms to increase their CSP? This line of inquiry could reveal whether firms, when evaluating CSP expenditures, are engaged in a rational pursuit or mere imitative behavior.

Despite these limitations, our study shows that different stakeholder groups have differing levels of salience for firm's financial performance. Results also indicate support for moderating influence of corporate governance structures. Lastly, our results show that commitment-based work practices help the firms improve their capacity to benefit from their performance on other CSP domains, which we believe indicates preliminary support for Barnett's (2007) concept of stakeholder influence capacity.

Annexure 1. Is CSP a Single Construct/ Index?

Factor analysis				
Method: principal factors		Number of obs		2072
Rotation: Orthogonal Varimax		Retained factors		3
		# of parameters		15
Factor	**Eigenvalue**	**Difference**	**Proportion**	**Cumulative**
Factor1	1.03208	0.25684	0.4072	0.4072
Factor2	0.77524	0.04798	0.3059	0.7131
Factor3	0.72726	.	0.2869	1
		Factor Loadings		
Variable	**Factor 1**	**Factor 2**	**Factor 3**	
Employee Relations (ER)	0.91	0.11	0.09	
Consumer Relations (CR)	0.08	-0.35	0.27	
Community relations (CMR)	0.07	0.55	0.19	
Environment (ENV)	-0.01	-0.03	0.63	
Diversity (DIV)	0.11	0.57	0.09	
Other concerns (OSP)	0.08	-0.13	0.27	
Cronbach's Alpha Analysis for Scale Measures				
Alpha: 0.2038				

Endnotes

1 This, of course, depends also on the specific context or the environment that a firm operates in.

2 Annexure 1 shows that even empirically it is more valid to study these components separately rather than as an index. The results of the factor analysis of our data show that these factors do not load together and are in fact distinct factors in their own right. The factor loadings and Eigenvalues are consistently low. In fact, the first two factors explain only about 70% of the variance. There are at least 3 different factors: employment relations, community relations, and environmental issues. We also used alpha to examine whether these dimensions can be used as a scale. The results in the bottom-half of Annexure 1 show that alpha is only 0.203. These results are robust to different model specifications and methods used for factor analysis, including rotation methods, and call into question the empirical validity of combining these distinct social performance domains into a social performance index/ scale.

3 We used normalization process to derive z-scores for standardizing/ centering all the variables to avoid influence of outliers.

4 Following Hillman and Keim (2001), we also use Market Value Added (MVA) in our sensitivity analysis, and the results mirror those in the use of MBR, where MVA refers to the difference between equity market valuation of the company and the capital (debt or equity) invested in the company. This is not surprising given the fact that the two measures essentially capture the same information, perhaps the only difference being that MBR is a ratio and MVA an absolute value.

5 We also examine the variance inflation factors (VIF) to detect multicollinearity; all VIF scores are below 2, and the mean VIF score is 1.2, well within the commonly specified rule of thumb of a score of 10 indicating multicollinearity problems (Cohen *et al.,* 2003). Thus, our results do not appear to suffer from multicollinearity.

6 Please refer to Oliver (1991) for a detailed discussion of firm responses and strategies for dealing with institutional pressures.

References

Ahmed AS, Duellman S (2007) Accounting conservatism and board of director characteristics: An empirical analysis. Journal of Accounting and Economics 43:411–437

Aldrich HE, Pfeffer J (1976) Environments of Organizations. Annual Reviews in Sociology 2:79–105

Alexander GJ, Buchholz RA (1978) Corporate Social Responsibility and Stock Market Performance. The Academy of Management Journal 21:479–486

Amato LH, Amato CH (2007) The Effects of Firm Size and Industry on Corporate Giving. Journal of Business Ethics 72:229–241

Amit R, Schoemaker PJH (1993) "Strategic assets and organizational rent". Strategic Management Journal, pp 33–46

Anderson EW, Sullivan MW (1993) "The antecedents and consequences of customer satisfaction for firms". Marketing Science 12(2):125–143

Aupperle KE, Carroll AB, Hatfield JD (1985) An Empirical Examination of the Relationship between Corporate Social Responsibility and Profitability. The Academy of Management Journal 28:446–463

Barnett ML (2007) Stakeholder Influence Capacity and the Variability Of Financial Returns To Corporate Social Responsibility. The Academy of Management Review (AMR) 32:794–816

Barney JB, Wright PM (1998) On becoming a strategic partner: The role of human resources in gaining competitive advantage. Human resource management 37:31–46

Barney JB (1991) "Firm Resources and Sustained Competitive Advantage". Journal of Management 17(1):99

Bebchuk LA, Cohen A (2005) "The Costs of Entrenched Boards". Journal of Financial Economics 78:409–33

Bennett RJ (1996) "Relationship formation and governance in consumer markets: transactional analysis versus the behaviourist approach". Journal of Marketing Management 12(5):417–436

Burke L, Logsdon JM, Mitchell W, Reiner M, Vogel D (1986) Corporate Community Involvement in the San Francisco Bay Area. California management review 28:122–141

Cyert RM, March JG (1963) "A behavioral theory of the firm, Englewood Cliffs", NY: Prentice Hall

Davis GF (1991) Agents without Principles? the Spread of the Poison Pill through the Intercorporate Network. Administrative Science Quarterly 36(4):583–613

Delaney JT, Huselid MA (1996) The Impact of Human Resource Management Practices on Perceptions of Organizational Performance. Academy of Management Journal 39:949–969

Delery JE, Doty DH (1996) Modes of theorizing in strategic human resource management: tests of universalistic, contingency, and configurational performance predictions. Academy of Management Journal 802–835

Dierickx I, Cool K (1989) "Asset Stock Accumulation and Sustainability of Competitive Advantage". Management Science 35(12):1504–1511

Dwyer FR, Schurr PH, Oh S (1987) "Developing buyer-seller relationships". The Journal of Marketing, pp 11–27

Fama EF (1998) Market efficiency, long-term returns, and behavioral finance. Journal of Financial Economics 49:283–306

Fey CF, Björkman I, Pavlovskaya A (2000) The Effect of Human Resource Management Practices on Firm Performance in Russia. The International Journal of Human Resource Management 11:1–18

Finkelstein S, Boyd BK (1998) How much does the CEO matter? The role of managerial discretion in the setting of CEO compensation. Academy of Management Journal 41:179–199

Fornell C, Mithas S, Morgeson FV III, Krishnan MS (2006) Customer Satisfaction and Stock Prices: High Returns, Low Risk. Journal of Marketing 70:3–14

George G (2005) Slack Resources And The Performance Of Privately Held Firms. The Academy of Management Journal 48:661–676

Gompers P, Ishii J, Metrick A (20039 Corporate governance and equity prices. The Quarterly Journal of Economics 118:107–155

Granovetter M (1985) "Economic Action and Social Structure: The Problem of Embeddedness". American Journal of Sociology 91(3):481

Gundlach GT, Murphy PE (1993) "Ethical and legal foundations of relational marketing exchanges". The Journal of Marketing, pp 35–46

Guthrie JP (2001) High-involvement Work Practices, Turnover, and Productivity: Evidence from New Zealand. Academy of Management Journal 44:180–190

Halme M, Niskanen J. (2001) Does Corporate Environmental Protection Increase or Decrease Shareholder Value? The Case of Environmental Investments. Business Strategy and the Environment 10:200–214

Hambrick DC, Werder A, Zajac EJ (2008) New Directions in Corporate Governance Research. Organization Science 19:381–385

Heskett JL (2008) The New Math of Customer Relationships, http://hbswk.hbs.edu/item/5884.html, Harvard Business School Working Knowledge, Cambridge, MA

Hill CWL, Phan P (1991) CEO tenure as a determinant of CEO pay. Academy of Management Journal 34:707–717

Hillman AJ, Keim GD (2001) Shareholder value, stakeholder management, and social issues: what's the bottom line? Strategic Management Journal 22:125–139

Hoffer GE, Pruitt SW, Reilly RJ (1988) The Impact of Product Recalls on the Wealth of Sellers: A Reexamination. The Journal of Political Economy 96:663–670

Hoskisson RE, Hitt MA, Johnson RA, Grossman W (2002) Conflicting voices: The effects of institutional ownership heterogeneity and internal governance on corporate innovation strategies, Academy of Management Journal 45:697–716

Huselid MA (1995) "The Impact of Human Resource Management Practices on Turnover, Productivity, and Corporate Financial Performance". The Academy of Management Journal 38(3):635–672

Johnson RA, Greening DW (1999) The Effects of Corporate Governance and Institutional Ownership Types on Corporate Social Performance. The Academy of Management Journal 42:564–576

King A, Lenox M (2002) Exploring the Locus of Profitable Pollution Reduction. Management Science 48:289–299

Laursen K, Foss NJ (2003) "New Human Resource Management Practices, Complementarities and the Impact on Innovation Performance". Cambridge Journal of Economics 27(2):243–263

Luo X, Bhattacharya CB (2006) Corporate Social Responsibility, Customer Satisfaction, and Market Value. Journal of Marketing 70:1–18

Maddala GS (2002) Introduction to Econometrics; 2nd eds (Prentice-Hall, New York, NY)

Mahon J, Griffin J (1999) "Painting a portrait", Business and Society 38(1):126–133

Malkiel BG (2003) The Efficient Market Hypothesis and Its Critics. Journal of Economic Perspectives 17:59–82

McGuire JB, Sundgren A, Schneeweis T (1988) Corporate Social Responsibility and Firm Financial Performance. The Academy of Management Journal 31:854–872

McKinsey & company (2002) Global Investor Opinion Survey: Key Findings

McWilliams A, Siegel D (2000) Corporate Social Responsibility and Financial Performance: Correlation or Misspecification? Strategic Management Journal 21:603–609

Morck R, Shleifer A, Vishny RW (1988) Management Ownership and Market Valuation: An Empirical Analysis. Journal of Financial Economics 20:293–315

Moskowitz M (1972) Choosing Socially Responsible Stocks. Business and Society Review 1:75

Oliver C (1991) "Strategic Responses to Institutional Processes". The Academy of Management Review 16(1):145–179

Orlitzky M, Schmidt FL, Rynes S (2003) Corporate Social and Financial Performance: A Meta-analysis, Organization Studies 24:403

Pfeffer J, Salancik GR (2003) The External Control of Organizations: A Resource Dependence Perspective. Stanford University Press, Palo Alto, CA

Pfeffer J (1972) "Size and Composition of Corporate Board of Directors: The Organization and its Environment". Administrative Science Quarterly 17:218–28

Pfeffer J, Salancik GR (1978) The External Control of Organizations: A Resource Dependence Perspective (Stanford University Press, Palo Alto, CA)

Reed D (2001) "Stalking the Elusive Business Case for Corporate Sustainability", World Resources Institute, Washington DC

Richard OC (2000) Racial Diversity, Business Strategy, and Firm Performance: A Resource-Based View. Academy of Management Journal 43:164–177

Richard OC, Murthi BPS, Ismail K (2007) The Impact of Racial Diversity on Intermediate and Long-Term Performance: The Moderating Role of Environmental Context. Strategic Management Journal 28:1213–1233

Russo MV, Fouts PA (1997) A resource-based perspective on corporate environmental performance and profitability. Academy of Management Journal 40:534–559

Seifert B, Morris SA, Bartkus BR (2004) "Having, Giving, and Getting: Slack Resources, Corporate Philanthropy, and Firm Financial Performance". Business & Society 43(2):135

Seifert B, Morris SA, Bartkus BR (2003) Comparing Big Givers and Small Givers: Financial Correlates of Corporate Philanthropy. Journal of Business Ethics 45:195–211

Sen S, Bhattacharya CB (2001) "Does Doing Good Always Lead to Doing Better? Consumer Reactions to Corporate Social Responsibility". Journal of Marketing Research 38(2):225–243

Solomon MR, Surprenant C, Czepiel JA, Gutman EG (1985) "A Role Theory Perspective on Dyadic Interactions: The Service Encounter". Journal of Marketing 49(1):99–111

Surroca J, Tribo JA (2006) Managerial Entrenchment And Corporate Social Performance

Thomas III, LG, Waring G (1999) Competing capitalisms: Capital investment in American, German, and Japanese firms. Strategic Management Journal 20:729–748

Turban DB, Greening DW (1997) Corporate social performance and organizational attractiveness to prospective employees. Academy of Management Journal 40:658–672

Ullmann AA (1985) Data in Search of a Theory: A Critical Examination of the Relationships among Social Performance, Social Disclosure, and Economic Performance of US Firms. The Academy of Management Review 10:540–557

Vance SC (1975) Are Socially Responsible Corporations Good Investment Risks? Management review 64:19–24

Waddock SA, Graves SB (1997) The Corporate Social Performance–Financial Performance Link, Strategic Management Journal 18:303–319

Wang J, Coffey BS (1992) Board composition and corporate philanthropy. Journal of Business Ethics 11:771–778

Woo CY, Willard GE, Daellenbach US (1992) Spin-off performance: A case of overstated expectations. Strategic Management Journal 13:433–447

Wood DJ, Jones RE (1995) Stakeholder mismatching: A theoretical problem in empirical research on corporate social performance. International Journal of Organizational Analysis 3:229–267

Wright PM, McMahan GC, McWilliams A (1994) "Human resources and sustained competitive advantage: a resource-based perspective". The International Journal of Human Resource Management 5(2):301–326

Wulf KD, G Odekerken-Schröder, D Iacobucci (2001) Investments in Consumer Relationships: A Cross-Country and Cross-Industry Exploration. Journal of Marketing 65:33–50

Zaltman G, L Zaltman (2008) Marketing Metaphoria: What Deep Metaphors Reveal About the Minds of Consumers (HBS Business Press, Cambridge, MA)

ZfB
ZEITSCHRIFT FÜR
BETRIEBSWIRTSCHAFT

The Importance of Social Consumerism Across Eight Countries

Patrice Auger, Timothy M. Devinney, Jordan J. Louviere

Abstract: This paper describes the results of several large empirical studies that investigated the impact of social product attributes on consumer purchase intentions. Our results show that some consumers are willing to pay for more socially acceptable products, but that most of those consumers do not know about the social product features of the products they purchase. Furthermore, our analyses demonstrate that consumers can be segmented based on their preferences for (or against) social product features and that these segments are not country specific and do not relate well to observable demographic factors.

Keywords: Social consumption · ethical consumerism · discrete choice modeling

JEL Classification: M14 · D11 · D12 · D46 · C25

Prof. Patrice Auger, Ph.D. (✉)
Melbourne Business School, University of Melbourne, Melbourne, Australia
E-mail: p.auger@mbs.edu

Prof. Timothy M. Devinney, Ph.D. (✉)
University of Technology, Sydney, Australia
E-mail: Timothy.Devinney@uts.edu.au

Prof. Jordan J. Louviere, Ph.D. (✉)
University of Technology, Sydney, Australia
E-mail: jordan.louviere@uts.edu.au

1 Introduction

Over the past decade there has been a growing debate about the importance of social consumerism (Auger et al. 2003, 2008; Bhattacharya/Sen 2004; Sen/Bhattacharya 2001). This debate has become increasingly complex since the 'ethical' component of social consumerism has evolved substantially from an almost exclusive focus on environmental issues (i.e., green consumerism) to a concept that more broadly incorporates matters of conscience and politics. These components can include a variety of issues such as animal welfare, labor standards, human rights, health-related issues, human rights and, of course, environmental issues (Carrigan et al. 2004; Crane 2001; Strong 1996).

In its broadest form, social consumerism can be defined as 'the conscious and deliberate choice to make certain consumption choices due to personal and moral beliefs' (Crane/Matten 2004). Hence, consumers can express their social concerns about products and organizations in two ways: 1) by choosing to purchase a product that meets certain social (or ethical) standards (e.g., not manufactured by children or not tested on animals), or 2) by choosing not purchase a product that fails to meet those criteria. The critical aspect of social consumerism from an organizational point of view thus centers on the purchasing decisions of consumers.

Recent survey based studies on social consumerism suggest that consumers are giving increasing consideration to the social components of products and business processes and that these concerns have financial implications for the businesses involved (Carrigan et al. 2004; Mason 2000; Rogers 1998; Uusitalo/Oksanen 2004). For example, a large-scale survey by MORI found that over one-third of consumers in the United Kingdom were seriously concerned with social issues (Mason 2000). The same survey also suggested that the potential for social products could be as high as 30 per cent of UK consumer markets. Similarly, a 2005 poll by Global Market Institute (GMI) across a wide range of countries including the US, UK, India, Australia, Canada and countries throughout Europe, found that 54 percent of consumers would be prepared to pay more for organic, environmentally friendly, or fair-trade products.

A related trend towards more consumer activism with respect to the 'social behaviors' of organizations can also be seen in recent, highly publicized developments such as the increasing number of large-scale protests directed at multinational corporations and international organizations. In fact, demonstrators have often become the main focus of news reports during large-scale meetings such as those of the WTO, World Bank, IMF, G7, UN and World Economic Forum. The number of groups focusing their attention on the social behaviors of companies also seems to have increased in recent years (Argenti 2004). For example, Elliott and Freeman (2001) identified over forty anti-sweatshop organizations in the US alone, a large number considering the focus of those organizations on a single issue and one that no doubt has increased in the subsequent years. Even larger numbers of groups and members can be found in areas such as environmental protectionism, human rights, and animal rights with such large and well-established organizations as Greenpeace, Amnesty International and the World Wildlife Fund.

The breadth of issues associated with social consumerism suggests that it has wide-ranging implications for managers and can affect the strategic decisions firms are making – from entering and selling into foreign markets to outsourcing work to specific countries.

For example, the importance consumers attach to the use of child labor in production could affect decisions about the use of offshore production, the selection of a country in which to produce, the human resources policies of sub-contractors, and/or whether or not to sell a particular product in a specific country market. Therefore, it is clear that businesses need to better understand the preferences of consumers with respect to the ethical and social features of products, especially when these businesses are involved internationally. Unfortunately, determining these preferences can be quite challenging given the relatively sensitive nature of the issues under consideration. The next section will highlight some of the main problems associated with research on social consumerism followed by a discussion of our research that focused on mitigating some of these problems (more detail can be found in Devinney et al. (2010)).

2 Problems with existing research on social consumerism

Despite the apparent importance of social consumerism, most of the conclusions from the studies mentioned above (and whether in the commercial or academic domains) are based on survey results that ask respondents to simply rank the importance of a list of social issues (Cotte 2009). They do not force consumers to trade-off social features of products against traditional utilitarian features such as brand or price, nor do they question the point at which one social cause takes precedent over more mundane aspects of life (such as interest rates or petrol prices). Hence, it is not unreasonable to believe that these surveys overstate the importance of social features, since there are clearly more socially acceptable answers (Auger/Devinney 2007). For example, few people would answer that they do not care about the use of child labor or the amount of pollution involved in the manufacturing of the products they consume when there is no cost in hiding their true preferences.

At the core of this problem is an issue referred to as 'incentive compatibility'– i.e., the extent to which the structure of the survey instrument allows or forces respondents to reveal their 'true' underlying behavior, preference or attitude (Carson et al. 2000). This is potentially a serious problem for research on social consumerism given the sensitivity of the issues under investigation, the lack of any penalty for not revealing the 'truth', and the obvious intent of the research to tap into social attitudes and behavior. In effect, incentive compatibility takes the form of a social desirability bias. As Schwarz (1999, p. 97) put it, "respondents may want to edit their private judgment before they report it to the researcher, due to reasons of social desirability and self-presentation".

Survey-based research on social consumerism also suffers from a number of other weaknesses that may affect the reliability and interpretability of the results. Two of the most relevant for our research are the features of rating scales and their informational content. The first issue is fairly obvious. The wording used in rating scales is critical and vague wording can lead to erroneous conclusions (Schwarz et al. 1998). For example, Uusitalo and Oksanen (2004) found that 70 percent of Finish respondents stated that a firm's business ethics had "some influence" on their purchase behavior. This relatively vague label begs for more information – how much "influence" is "some influence" and is one person's "some influence" equivalent to another person's "some influence"?

The second issue, the informational content of the rating scale, is much less obvious. In short, rating scales (and more generally questionnaires) are often an important source of information for respondents. That is, respondents draw on the information embedded in rating scales (and questionnaires) to arrive at an answer (Schwarz 1999). Furthermore, research also shows that respondents utilize this information more heavily when the behavior is poorly represented in memory and/or when the behavior is ill defined, which are clearly the case when asking respondents about ethical and social issues.

As a result, several researchers have proposed that a lack of incentive compatibility in traditional survey instruments has led to an attitude-behavior gap with respect to the impact of social issues on consumer purchases (Boulstridge/Carrigan 2000; Carrigan/Attala 2001; Ulrich/Sarasin 1995). That is, consumers indicate in surveys that social issues are important, but do not change their purchase behaviors accordingly. Using experimental methodologies, primarily choice experiments and best-worst scaling, we have attempted to deal with the aforementioned problems and gain a better understanding of consumers' "true preferences" with respect to social and ethical issues. The next section will describe our research as well as the main results and implications of two of our large-scale empirical studies.

3 The research projects

Over the past several years, we have conducted a number of large-scale empirical studies with several hundred consumers across a variety of countries (Devinney et al. 2010). Our main objectives have been to answer a series of questions with respect to social consumerism:

1. Do consumers know about the social features of the products they purchase?
2. Are there groups of consumers who are willing to pay more for socially desirable products?
3. If so, how much are they willing to pay for those products?
4. Are there differences between different country markets in answers to the previous three questions?

In addition to beginning a dialogue around these very fundamental issues, our work has the added benefit of introducing to the field rigorous experimental methodologies, in the form of choice experiments and best-worst scaling experiments, to more effectively deal with some of the problems of the survey methods outlined previously.

Choice experiments enabled us to more accurately answer our questions since they force individuals to trade-off the social features of products against more traditional features (Auger et al. 2003, 2008; Auger/Devinney 2007; Louviere et al. 2000). This provides a more realistic simulation of the purchase process since consumers are forced to make such trade-offs (usually specific features against the price of the product) when they make actual purchases. In effect, we treated social features (e.g., the use of child labor or poor working conditions) as any other product feature and included them as part of a set of features that described specific products (see Appendix A for an example of a choice experiment).

Our experiments were conducted in two large-scale studies with over 1,000 respondents from eight countries: Australia, Hong Kong, Germany, Spain, Turkey, India, South Korea, and the United States. We specifically chose those countries in an attempt to maximize differences in culture, level of economic development, geographic location, and religious beliefs (Auger et al. 2010). Those, and other demographic variables such as age, education, and gender, have been shown to impact social beliefs in previous research (Al-Khatib et al. 2005; Al-Khatib et al. 1997; Rawwas 2001; Rawwas et al. 1998).

Each experiment required respondents to complete three sections: 1) evaluate their most recently purchased brand; 2) decide whether to purchase various profiles of products (either athletic shoes, bath soap, AA batteries, or laundry detergent); and 3) answer a series of socio-demographic questions.

We gathered data for four types of products: AA batteries, athletic shoes, bath soap, and laundry detergent. We selected these products for the following reasons. First, they enabled us to investigate the importance of different sets of social issues, namely a variety of environmental issues for batteries, bath soap, and laundry detergent and labor issues for athletic shoes. Second, the products were familiar to and purchased by most consumers in our samples including consumers from developing countries. Knowledge of the product categories and prior purchase experience were important since we also asked respondents to tell us about the attributes of their most recently purchased products. Prior purchase experience also facilitated the experimental tasks since respondents already understood the nature of the product attributes. Third, the products also differed in their level of consumer involvement in the purchase process. Specifically, athletic shoes are considered high involvement products compared to the other products since consumer search is more intensive and the price more noticeable to the consumer. In this paper, we focus our attention almost exclusively on athletic shoes since it is the only product category for which we have data for all respondents across all countries. We alternated the other three products randomly among our respondents so as to keep the time required to complete the experiments manageable.

The best-worst scaling (BWS) experiment was done only in our second study – which excludes Australia and Hong Kong – and focused on a broader set of social issues. We selected the best-worst scaling method since it could effectively overcome a variety of problems in the use of rating scales across countries. BWS is a multiple-choice extension of the paired comparison approach that is scale-free and forces respondents to make a discriminating choice among the issues under consideration. As Finn and Louviere demonstrated (1992, p. 13), "BWS models the cognitive process by which respondents repeatedly choose the two objects in varying sets of three or more objects that they feel exhibit the largest perceptual difference on an underlying continuum of interest".

In a best-worst (BW) study, the underlying continuum is the respondents' degree of concern about a set of objects, in our case social and ethical issues. Specifically, the BW experiment examined sixteen social and ethical issues (see Appendix B for a list of issues and for an example of a BW questionnaire). The experiment required each individual to examine twenty sets of four issues and indicate which issue of the four they considered "most important" and which they considered "least important". As such, there is a significantly reduced likelihood that we would encounter bias in the use of a rating scale since there is only one way to choose something as most (or least) important (Cohen/

Neira 2003). That is, BWS eliminates differences in the way that human subjects use rating scales, including cultural differences in rating styles if they exist. The reader is directed to Marley and Louviere (2005) for a more complete description of the scale properties of the approach and Auger et al. (2007) for a complete description of the work discussed here.

4 The main results and implications

Our results over the two studies are fairly consistent. First, we found that the vast majority of consumers do not know about the social features of the products they buy. Figure 1 presents of summary of the most recent purchase results for athletic shoes across samples and countries. The results show that over 80 per cent of consumers across our studies did not know about the social features of their most recently purchased athletic shoes. Conversely, the vast majority (almost 90 percent overall) could remember the functional features of these same shoes. We found a similar pattern across the other three product categories with the exception of AA batteries for which consumers possessed much better knowledge of the social features. This was still a significant gap between knowledge of functional and social features for batteries, but the gap was much smaller.

These were not surprising results given that companies do not publicize ethical transgressions. Hence, it is clear that part of the attitude-behavior gap when it comes to social consumerism may be explained by a lack of knowledge about products. That is, consumers may actually care about social features of the products they purchase, but simply don't have the product-specific information required to make more informed decisions.

Second, our results also show that there are groups of consumers who are willing to pay for more socially acceptable products. For example, figure 2 presents willingness-to-pay estimates for all the attributes of shoes (both functional and social) for our three samples of respondents in our first study. The figure shows that some consumers were willing to pay a significant amount for more socially desirable athletic shoes, especially for products

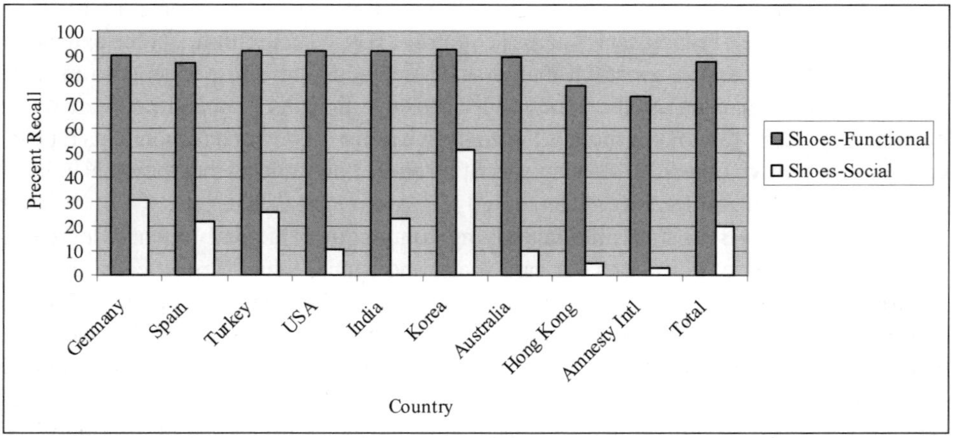

Fig. 1. Knowledge of Most Recent Purchase of Athletic Shoes by Sample

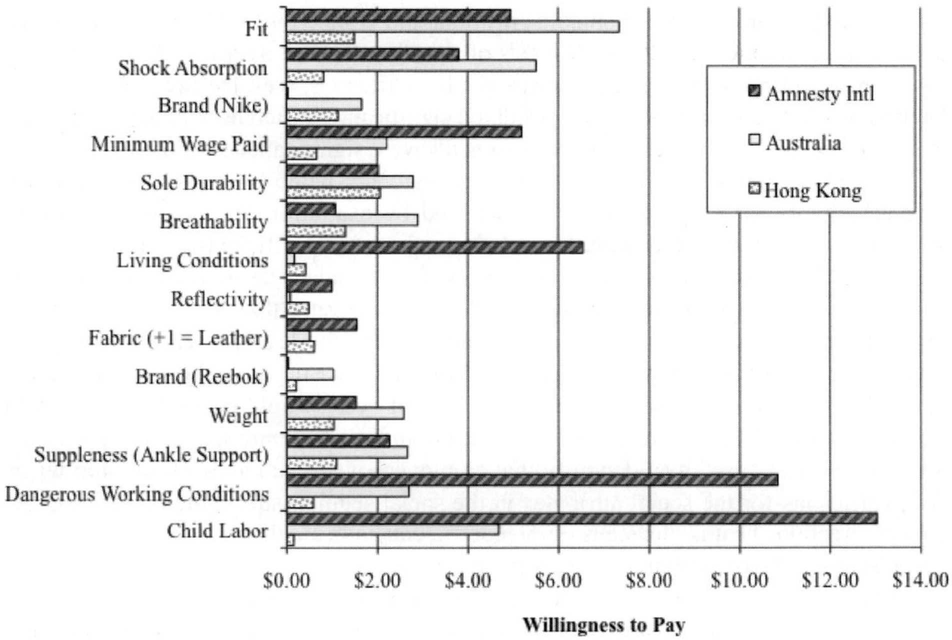

Fig. 2. Willingness-to-Pay Estimates for Athletic Shoes – First Study

not manufactured by children and made under safe working conditions. The figure also reveals, as we expected, that Amnesty International supporters were willing to pay more for socially desirable products than respondents from either Hong Kong or Australia (Amnesty International supporters were also based in Australia). What is more interesting is that consumers were not willing to sacrifice functional features of products for social ones. That is, some consumers were willing to pay more for more socially desirable products, but only if the products met minimum standards with respect to functional features. We should also qualify these results in that they indicate that some individuals will pay something for social features in ideal circumstances. Hence, our results are an upper bound.

Third, choice experiments also enabled us to segment (or group) consumers based on their preferences. We used a relatively sophisticated type of regression analysis referred to as latent class regression analysis (LCRA) to accomplish this task (Andrews/Currim 2003a; Andrews/Currim 2003b; Kamakura/Russell 1989). LCRA allowed for the classification of individuals into segments (often called classes) and develops regression models for each of the segments simultaneously. These segments are referred to as latent segments since their formation does not depend on a group of pre-specified clustering variables (as is the case in traditional clustering methods). Instead, the latent segments are formed with discrete unobserved variables, which greatly improves the ability of researchers to identify meaningful segments in circumstances where observed variables (e.g., socio-demographics) have proven to be ineffective.

Figure 3 presents the standardized coefficients for LCRAs of product attributes on choice (i.e., the decision of whether or not to purchase a specific product) for athletic shoes for our second study (again excluding Australia and Hong Kong). We also included six markers to indicate commonly-used levels of significance (i.e., p=.05; p=.01; p=.001) for easier interpretation of the results. Basically, all coefficients, with the exception of price, beyond the first marker (p=.05) are considered significantly different from zero (for price, we show price elasticity and all price coefficients were significant at the .05 level).

The analysis clearly identified three distinct segments, namely brand, price and social. The descriptors for the segments were selected by examining the dominant set(s) of attributes (i.e., the attributes with the largest standardized coefficients) within each segment. It is important to note that the wording for the social attributes is a mixture of "positive" and "negative" statements. We did this to ensure that respondents took the experimental task seriously and paid close attention to the levels of each attribute. As such, some of the coefficients are positive while others are negative. For example, child labor is strongly negative since an attribute level of "yes" would mean that the product was manufactured by children. Conversely, minimum wage is strongly positive since an attribute level of "yes" would signify that employees were paid above minimum wage. All coefficients for the social attributes in the social segment are in the correct and expected direction. That is, the signs of the coefficients indicate that the respondents in the social segment favored products that were more "socially desirable".

Similar to the results from our first study, figure 3 shows that the functional attributes, including brand and price, are not irrelevant to respondents in the social segment. For example, respondents in the social segment for shoes still show a fair amount of price sensitivity (much greater than for respondents in the brand segment) and a preference towards alternative brands (i.e., not well-known brands such as Nike and Adidas), among others. What these results suggest is that managers cannot simply ignore the core functional attributes of their products to create more socially acceptable ones. In other words,

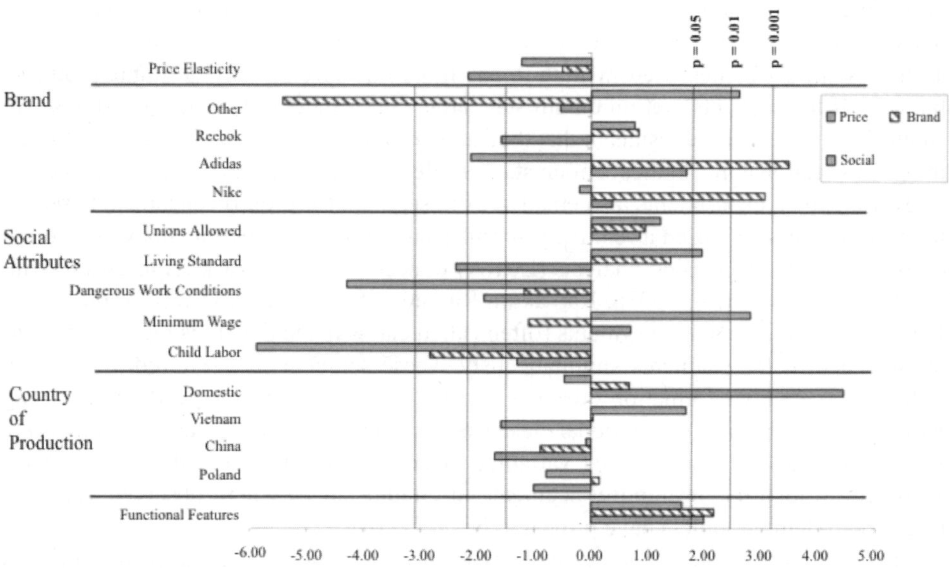

Fig. 3. Segments of Socially-Conscious Consumers for Athletic Shoes

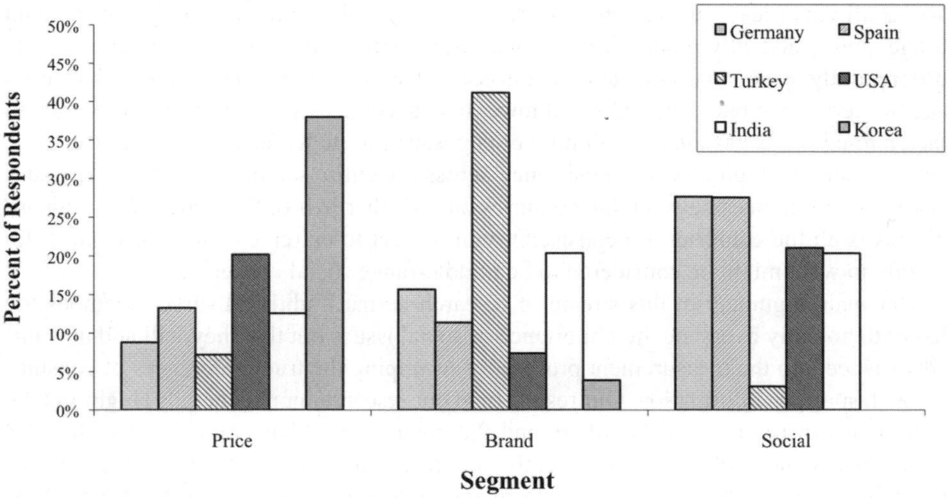

Fig. 4. Country Membership by Segment for Athletic Shoes

consumers do not appear willing to sacrifice functionality for social desirability. What these consumers are telling us is that they purchase products to fill a certain basic set of needs and that no amount of social desirability is likely to compensate for a failure to meet these basic needs.

Interestingly the segments derived were also not country-specific. That is, membership in a particular segment was only weakly related to the country of residence of the respondent. Figure 4 presents country memberships by segment for athletic shoes. The results show that all countries are represented in all three segments with the exception of Korean respondents in the social segment. What we have consistently found across all of our studies is that there is considerably more variation in preferences within countries than between countries. What this implies is that rough segmentation approaches based on aggregate demographics must be approached with extreme caution and the natural tendency in cross-cultural research to seek out differences can be biased.

In fact, what has been most surprising from our research is what we did not find. As just stated, we found few differences between the different country markets with respect to the importance of social consumerism. This is not to suggest that there were no differences at all, but that the differences were much smaller than expected or reported in previous research. Furthermore, our results also showed few differences based on typical demographic variables such as age, gender, education, and income. This is even more surprising given that we had specifically selected countries to maximize differences among those variables.

The lack of cross-country differences is best exemplified by results from our BWS experiments. Figure 5 presents the results across the six countries in our sample. The results are quite revealing with respect to the lack of differences between countries. First, four issues clearly received consistently high ratings across all countries – human rights, child labor (although Koreans give it a neutral rating), safe working conditions, and good living conditions. Second, four other issues clearly and consistently received low ratings across all countries – recycled packaging, use of animal by-products, recycled material usage, and genetically modified materials. The remaining eight issues are viewed quite differentially by country. For example, the issue of animal rights is rated higher than average by Germany, Spain, and India, but lower than average by USA, Turkey, and Korea. In fact, animal rights has the lowest rating of any issue for the Korean sample. Hence, these analyses show a high level of consistency across the entire sample for half of the issues and some differences between the countries on the other half of the issues. What this indicates is all the countries are consistent with respect to extreme evaluations and differ mainly in what might be considered to be middle-range social concerns.

Our main argument in this stream of research is that traditional survey methods are likely to not only overstate the importance of social issues but that they will add unwanted variance into the measurement process by swamping the true preferences of consumers with spurious information. Our results (and our research more generally) highlight the importance of rigorous methodology and the potential problems associated with traditional survey methods. More importantly, this research demonstrates the importance of incentive compatibility within the field of social consumerism and the tendency of respondents to answer questions about social issues in a socially acceptable way. Though our research has focused primarily on social consumerism, its implications are thus relevant for anyone conducting and/or interpreting research that deals with sensitive issues.

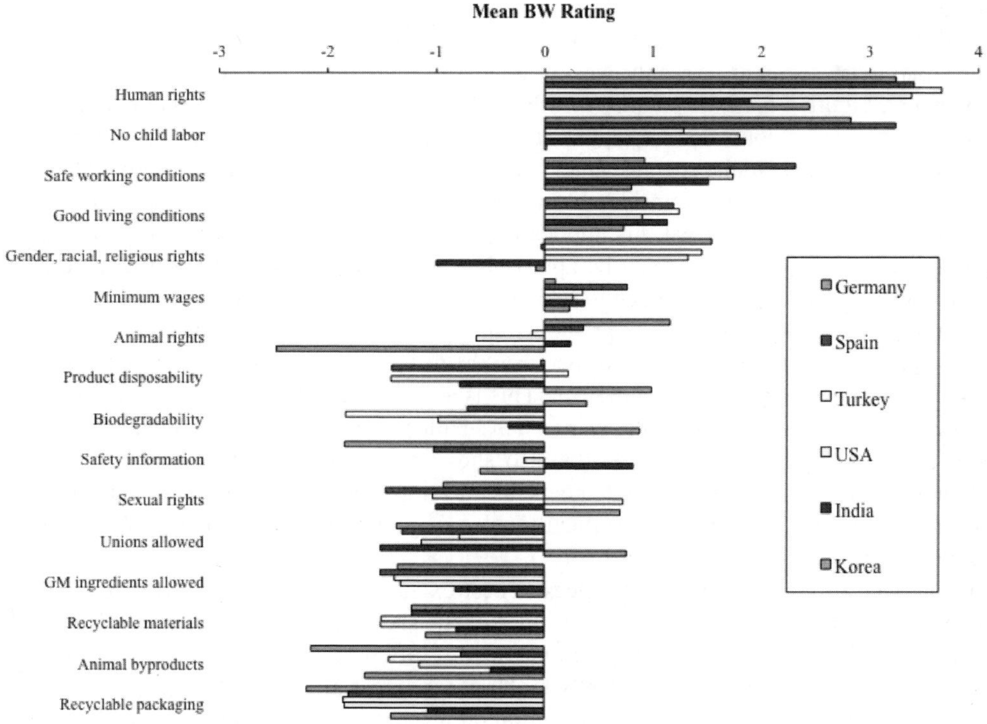

Fig. 5. Results of Best-Worst Scaling Experiment Across Countries

Appendix A: Hypothetical Product Profile for Athletic Shoes

Features of the Shoes	Features of Shoe #8
Shock absorption/cushioning	High
Weight	Heavier
Ankle support	Low cut
Sole durability	Short
Breathability/ventilation	High
Fabrication materials	Synthetic
Reflectivity at night	No
Comfort/fit	Low
Is child labor used in making the product?	No
Are workers paid above minimum wage?	Yes
Are workers' working conditions dangerous?	No
Are workers' living conditions at the factory acceptable?	No
Are workers allowed to unionize?	Yes
Country of production	Vietnam
Brand of shoes	Nike
Price	$130
1. If the shoes described above were available in your local shops now, would you consider trying it (Tick ONE box only)? ☐ No ☐ Yes	
2. If the shoes described above were available in your local shops now, would you buy it instead of or in addition to your current shoes next time you shop for shoes (Tick ONE box only)? ☐ No ☐ Yes	

Appendix B: Best-Worst Experiment

In this section, we will present you with sixteen social and ethical issues. These will be organized in groups of four over the next two pages (a total of twenty groups or questions). For each group, select the **one issue** among the four that is **least important** to you and the **one issue** that is **most important** to you. Please make sure that you select only one least important and one most important for each group of four issues. We have included a description of the issues below; please keep them in mind throughout the rest of this section.

- **Animal rights** – describes the general treatment of animals for commercial purposes such as the use of animals for product testing, the displacement or killing of animals for natural resource exploitation (e.g., logging), or the cruel use of animals for entertainment.
- **Animal byproducts used** – Indicates that the product is made using animal byproducts such as animal fat or lard.
- **Product biodegradability** – indicates that the materials used to make a product can be broken down naturally and hence are safer for the environment.
- **Products made from recyclables** – indicates that some or all of the materials used to make a product were obtained from recycled sources.
- **Product safety information provided** – means that information about the safe use of a product and/or potential dangers from using a product is/are included with the product.
- **Human rights** – describes the basic rights of all people as stated in the Universal Declaration of Human Rights such as the right to food, clothing, housing, education, etc.
- **Packaging recyclability** – indicates that part or all packaging materials can be recycled for future use (e.g., product packages, food containers, shipping boxes, etc.).
- **Product disposability** – indicates that a product can be disposed of without causing undue damage to the environment.
- **Paying minimum wages** – signifies that companies adhere to the minimum wage standards of the country (ies) in which they are operating.
- **Unions allowed** – indicates that unionization is legal within a country and that companies producing in that country do not attempt to prevent or curtail the unionization of their workers.
- **Minimum living conditions met** – means that companies supply their employees with basic and acceptable living accommodations when required.
- **Sexual rights** – indicates that discrimination against individuals based on their sexual orientation is not allowed.
- **Safe working conditions** – signifies that companies follow a set of procedures to create a safe working environment for their workers.
- **Child labor not used** – means that companies do not use workers under the minimum working age in the country (ies) in which they are operating.
- **Genetically modified material used** – indicates that the use of genetically modified (GM) materials is allowed within a country and that companies use GM materials in their products.
- **Gender, religious, racial rights** – indicates that discrimination based on gender, religion, or race is not allowed.

– EXAMPLE –

In this example, sexual rights are least important and human rights are most important. Please notice that only **one issue** was selected in **each column** (**Least Important** and **Most Important**).

	Least Important		Most Important
A	☐ Only ⇨ √ ☐ ☐	Animal rights Sexual rights Human rights Gender, religious, racial rights	☐ ☐ √ ⇦ Only ☐

Question No.	**Which issue matters LEAST to you? (<u>tick ONLY ONE box for each question</u>)**	**Sets of social and social issues for you to consider**	**Which issue matters MOST to you? (<u>tick ONLY ONE box for each question</u>)**
1	☐ ☐ ☐ ☐	Animal rights Product biodegradability Products made from recyclables Product safety information provided	☐ ☐ ☐ ☐
2	☐ ☐ ☐ ☐	Human rights Product biodegradability Packaging recyclability Product disposability	☐ ☐ ☐ ☐
3	☐ ☐ ☐ ☐	Product biodegradability Paying minimum wages Unions allowed Minimum living conditions met	☐ ☐ ☐ ☐
4	☐ ☐ ☐ ☐	Sexual rights Product biodegradability Safe working conditions Animal byproducts used	☐ ☐ ☐ ☐
5	☐ ☐ ☐ ☐	Gender, religious, racial rights Product biodegradability Child labor not used Genetically modified material used	☐ ☐ ☐ ☐
6	☐ ☐ ☐ ☐	Human rights Animal rights Gender, religious, racial rights Sexual rights	☐ ☐ ☐ ☐
7	☐ ☐ ☐ ☐	Animal rights Packaging recyclability Paying minimum wages Safe working conditions	☐ ☐ ☐ ☐
8	☐ ☐ ☐ ☐	Animal rights Product disposability Genetically modified material used Minimum living conditions met	☐ ☐ ☐ ☐

References

Al-Khatib JA, Stanton AD, Rawwas MYA (2005) "Ethical Segmentation of Consumers in Developing Countries: A Comparative Analysis". International Marketing Review 22(2):225–46

Al-Khatib JA, Vitell SJ, Rawwas MYA (1997) "Consumer Ethics: A Cross-Cultural Investigation". European Journal of Marketing 31(11/12):750–67

Andrews RL, Currim IS (2003a) "A Comparison of Segment Retention Criteria for Finite Mixture Logit Models". Journal of Marketing Research 40(2):235–43

— (2003b) "Retention of Latent Segments in Regression-Based Marketing Models". International Journal of Research in Marketing 20(4):315–21

Argenti PA (2004) "Collaborating with Activists: How Starbucks Works with NGOs". California Management Review 47(1):91–116

Auger P, Burke PF, Devinney TM, Louviere JJ (2003) "What will Consumers Pay for Social Product Features?" Journal of Business Ethics 42(3):281–304

Auger P, Devinney TM, Louviere JJ, Burke PF (2008) "Do Social Product Features have Value to Consumers?" International Journal of Research in Marketing 25(3):183–191

Auger P, Devinney TM, Louviere JJ, Burke PF (2010) "The importance of social product attributes in consumer purchasing decisions: A multi-country comparative study". International Business Review, (in press)

Auger P, Devinney TM (2007) "Do What Consumers Say Matter? The Misalignment of Preferences with Unconstrained Ethical Intentions". Journal of Business Ethics 76(4):361–383

Bhattacharya CB, Sen S (2004) "Doing Better at Doing Good: When, Why, and How Consumers Respond to Corporate Social Initiatives". California Management Review, 47(1):9–24

Boulstridge E, Carrigan M (2000) "Do Conusmers Really Care about Corporate Responsibility? Highlighting the Attitude-Behavior Gap". Journal of Communication Management 4(4):355–68

Carrigan M, Attala A (2001) "The Myth of the Ethical Consumer-Do Ethics Matter in Purchase Behaviour?" Journal of Consumer Marketing 18(7):560–77

Carrigan M, Szmigin I, Wright J (2004) "Shopping for a Better World? An Interpretive Study of the Potential for Ethical Consumption Within the Older Market". Journal of Consumer Marketing 21(6):401–17

Carson RT, Groves T, Machina MJ (2000) "Incentive and Informational Properties of Preference Questions". Unpublished Working Paper, University of California, San Diego

Cohen SH, Neira L (2003) "Measuring Preference for Product Benefits Across Countries: Overcoming Scale Usage Bias with Maximum Difference Scaling," in ESOMAR 2003 Latin America Conference Vol. 266. Punta del Este, Uruguay: ESOMAR

Cottee J (2009) "Socially Conscious Consumers: A Knowledge Project for the RNBS." Unpublished working paper, Ivey Business School

Crane A (2001) "Unpacking the Ethical Product". Journal of Business Ethics 30(4):361–73

Crane A, Matten D (2004) Business Ethics: A European Perspective: Managing Corporate Citizenship and Sustainability in the Age of Globalization. Oxford: Oxford University Press

Devinney TM, Auger Pat, Eckhardt GM (2010) The Myth of the Ethical Consumer. Cambridge, UK: Cambridge University Press

Elliott KA, Freeman RB (2001) "White Hats or Don Quixotes? Human Rights Vigilantes in the Global Economy". Cambridge, MA: National Bureau of Economic Research

Finn A, Louviere JJ (1992) "Determining the Appropriate Response to Evidence of Public Concern: The Case of Food Safety". Journal of Public Policy and Marketing 11(1):12–25

Kamakura Wagner A, Russell GL (1989) "A Probabilistic Choice Model for Market Segmentation and Elasticity Structure". Journal of Marketing Research 26(4):379–80

Louviere JJ, Hensher DA, Swait JD (2000) Stated Choice Methods: Analysis and Applications. Cambridge, UK: Cambridge University Press

Mason T (2000) "The Importance of Being Ethical". Marketing, 26 October, 27

Rawwas MYA, Patzer GL, Vitell Scoff J (1998) "A Cross-Cultural Investigation of the Ethical Values of Consumers: The Potential Effect of War and Civil Disruption". Journal of Business Ethics 17(4):435–48

Rawwas MYA (2001) "Culture, Personality and Morality: A Typology of International Consumers' Ethical Beliefs". International Marketing Review 18(2):188–209

Rogers D (1998) "Ethical Tactics Arouse Public Doubt," Marketing (August 6), pp 12–13

Schwarz, Norbert (1999), "Self-Reports: How the Questions Shape the Answers". American Psychologist 54(2):93–105

Schwarz N, Grayson CE, Knauper B (1998) "Formal Features of Rating Scales and the Interpretation of Question Meaning". International Journal of Public Opinion Research 10(2):177–83
Sen S, Bhattacharya CB (2001) "Does Doing Good Always Lead to Doing Better? Consumer Reactions to Corporate Social Responsibility". Journal of Marketing Research 38(No. May):225–43
Strong C (1996) "Features Contributing to the Growth of Ethical Consumerism-A Preliminary Investigation". Marketing Intelligence & Planning 14(5):5–13
Ulrich P, Sarasin Charles (1995) Facing Public Interest: The Ethical Challenge to Business Policy and Corporate Communications. Boston: Kluwer Academic Publishers
Uusitalo O, Oksanen R (2004) "Ethical Consumerism: A View from Finland". International Journal of Consumer Studies 28(3):214–21

ZfB
ZEITSCHRIFT FÜR BETRIEBSWIRTSCHAFT

Corporate Social Responsibility in Chinese Supplier Firms

Anna-Maria Schneider, Anja Schwerk

Abstract: Corporations with global supplier networks use Codes of Conduct in their endeavor to improve the working conditions in the supplier firms, above all in emerging markets. Regular audits should support the compliance with the codes. Practice, however, shows that in many cases Codes of Conduct are not implemented by the supplier firms. The present article analyzes the process of CSR realization in supplier firms using an extensive literature survey as well as semi-structured interviews. The results support the underlying assumption that a standard-based CSR approach without the creation of CSR awareness in the supplier firms has little prospect of success and may lead to non-compliance of the suppliers. The step-by-step implementation of CSR measures in combination with dialogue, exchange of experience, trainings and incentive and control mechanisms will increase the degree to which CSR will be realized in the supplier firms.

Keywords: Corporate social responsibility · code of conduct · supply chain · China · challenges

JEL Classification: M14

Anna-Maria Schneider (✉)
Institute of Management, Humboldt-Universität zu Berlin, Rosenstr. 19, 10178 Berlin, Germany
E-mail: anna-maria.schneider@wiwi.hu-berlin.de, URL: http://lehre.wiwi.hu-berlin.de/Professuren/bwl/management

Dr. Anja Schwerk (✉)
Institute of Management, Humboldt-Universität zu Berlin, Rosenstr. 19, 10178 Berlin, Germany
E-mail: schwerk@wiwi.hu-berlin.de, URL: http://lehre.wiwi.hu-berlin.de/Professuren/bwl/management

1 Introduction

In recent years, academic and corporate interest regarding corporate social responsibility (CSR) in the supply chain has risen considerably. The nature of business and the complexity of doing business are changing from firms which manufacture products within wholly owned facilities to global supply chain networks. Therefore, the concept of CSR is no longer an internal and therefore limited corporate issue, but encompasses the entire supplier network (Andersen/ Skjøett-Larsen 2009).

Firms are increasingly active in global markets, as consumers, suppliers and production plants are spread all over the world. In this context, China has become the largest producing nation and the second largest exporter in the world. Moreover, Chinese supplier firms are integrated in most global supply chains.

Within global supply chains, large Western companies usually operate as focal firms and are the central actors (Seuring/Müller 2008). By applying external pressure, stakeholder groups such as non-governmental organizations (NGOs), buyers, and governments insist that the companies take responsibility for ensuring humane working conditions as well as the containing of the environmental impact of production plants, notably in emerging and developing economies.

In this context, codes of conduct (CoC) are a popular tool for CSR realization which is used by focal firms to manage and monitor their suppliers' ethical and socially responsible practices (Waddock et al. 2002). Codes of conduct define required behaviors, responsibilities, and actions expected from suppliers. Common codes of conduct include rules concerning working hours and days, overtime per month, guaranteed minimum wages and overtime compensation, health and safety measures and training, and child labor. In some cases there are specifications on physical examinations, employment contracts as well as accident and pension insurance (Egels-Zandén 2007). CoC are a "weak form of regulation", which means that they are rarely legally enforceable, but often are incorporated into the contract with the supplier (Frenkel 2004).

Western companies impose CoC on their Chinese suppliers which are often based on general environmental and social standards. Chinese suppliers have to comply with these standards, which might be documented by implementing related management systems for environmental (e.g. ISO 14001, EMAS) and social (SA 8000) issues (Seuring/Müller 2008). However, it has been observed that in many cases, Chinese suppliers do not comply with CoC, environmental or social standards. The reasons for this are manifold: Some suppliers are simply not able to follow the requirements due to the related costs or the absence of management competencies. Others consciously breach the CoC (Jiang 2009).

Therefore, the general framework for CSR in Chinese supplier firms has to be analyzed. What might be reasons for resistance and non-compliance regarding CoC and standards? The question arises whether imposed instruments such as CoC and standards can actually address the prevalent challenges in China. At the same time, deriving options for managerial action to successfully implement CoC and standards in Chinese supplier firms is of major interest.

The overall thesis of this paper is that the implementation of stand-alone CoC is not the appropriate measure to support CSR in China. Rather, it might be necessary to create awareness and to explain the benefits of CSR to Chinese suppliers. Thus, the solution may not be a purely standard-based approach, but a more dialogue-oriented way.

2 CSR in supply chains

Spence and Bourlakis (2009, p. 291) define CSR in the supply chain as "the chain-wide consideration of, and response to, issues beyond the narrow economic, technical and legal requirements of the supply chain to accomplish social (and environmental) benefits along with the traditional economic gains which every member in that supply chain seeks".

Meaning and focus of CSR differ between countries and depend on underlying cultural, economic, legal, environmental as well as social conditions (Baughn et al. 2007). In emerging nations such as China, CSR measures rather imply compliance with human rights and existing laws by supplier firms than voluntary activities beyond legal requirements. For instance, China has a well elaborated labor law; working hours are not allowed to exceed eight hours a day and 44 hours a week on average (Labor Law of the People's Republic of China 1995, Ch. 4, Art. 36). In the daily routine these requirements are often not fulfilled by the supplier firms. CSR engagement in Chinese supplier firms is at an elementary stage and focuses on legal compliance (Lam 2009). CSR instruments such as standards and codes of conduct aim to support the compliance with existing requirements and laws.

Generally, the scientific literature on CSR in supply chain management is diversified and different aspects are accentuated. Most scientific contributions still focus on environmental issues (Seuring/Müller 2008). Concepts such as green supply chain management or sustainable supply chain management include a holistic system perspective (Andersen/Skjøett-Larsen 2009). Impacts of the entire supply chain on the environment and resources are considered (Sarkis 2003). Walker et al. (2008) focus on drivers and barriers for environmental supply chain management. Kovács (2008) studies the criticality of the supplier which influences the environmental requirements. Non-critical suppliers are faced with little environmental requirements. On the other hand, critical or strategic suppliers have to meet higher environmental standards (Kovács 2008). Cruz and Wakolbinger (2008) go further by developing a framework for the analysis of optimal levels of CSR engagement in a multi-period supply chain network.

Since the 1990s, social aspects in supply chain management have been studied increasingly (e.g. Carter/Jennings 2002; Davies/Crane 2003). For example, Carter and Jennings (2002) detect that purchasing social responsibility has a positive impact on supplier relationships as well as on the social performance of the supplier firms. Hutchins and Sutherland (2008) identify indicators to measure social sustainability in supply chains. They classify five dimensions of social sustainability: equity, health, education, housing security, and population. According to this classification, indicators are ratios such as the ratio of average female wage to male wage or adult literacy rate (Hutchins/Sutherland 2008).

Some authors focus on labor conditions in global supply chains (Yu 2008; Kortelainen 2008; Van Tulder et al. 2009) or on special issues such as working conditions and wages of female workers (Prieto-Carrón 2008). Zutshi et al. (2009) highlight the topic of child labor in global organized supply chains. In particular, supplier firms in emerging and developing countries such as China, India, Philippines, Malaysia, Indonesia, Myanmar, Bangladesh, and Thailand exploit child labor within the production process.

The question arises how CSR can be realized in global supplier networks. Following Waddock et al. (2002), codes of conduct are the most popular tool to improve working conditions and to support the compliance with social and environmental standards. In the scientific literature, especially codes of conduct are increasingly discussed (e.g. Egels-Zandén/Wahlqvist 2007; Bondy et al. 2008; Yu 2008; Jiang 2009). Van Tulder et al. (2009) conclude that the involvement of stakeholders in the design process increases the acceptance and probability of implementation of CoC.

Frenkel (2004) found that the nature of relations between a buying firm and its contract suppliers shapes the way CoC are implemented. Therefore, the supplier compliance will be higher when the buyer company prefers to develop a long-term relationship where the contractor has an ongoing incentive to work with the company to continuously improve productivity and quality. The opposite of such a relational contracting arrangement is a short-term transactional relationship where the contractor is not a key or strategic supplier for the buyer company.

Cramer (2008) points out that little is known about how firms can successfully organize and control appropriate CSR practices in their supply chain. Therefore, Cramer (2008) provides a step-by-step plan as guidance to firms for choosing appropriate CSR measures (see Fig. 1). The process of organizing CSR depends on the particularities of the specific

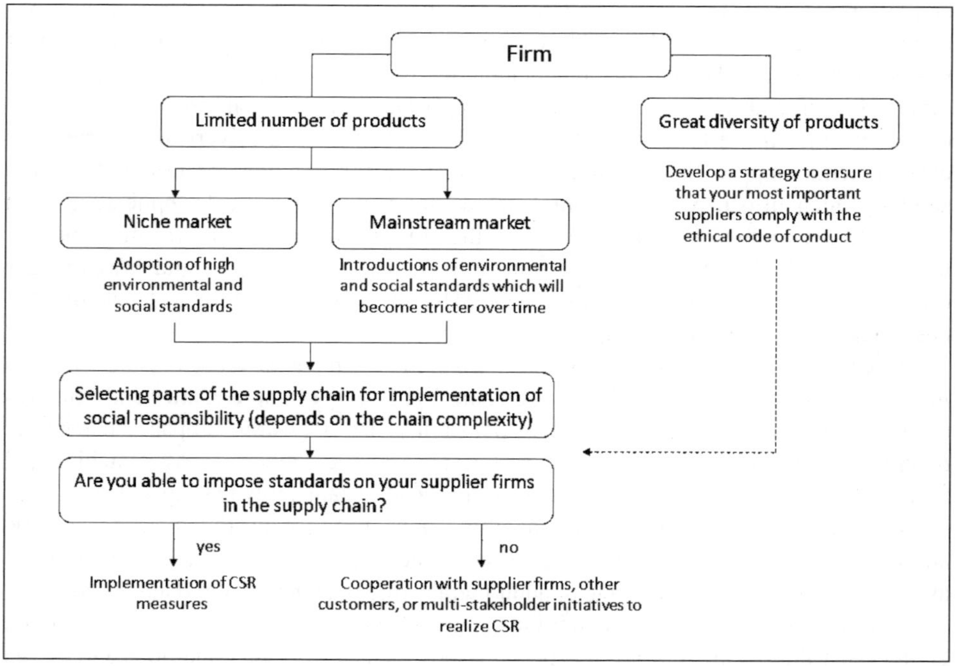

Fig. 1. Step-by-step plan for global chain responsibility according to Cramer (2008)

supply chain relationship. Additionally, the characteristics of the supplier, the company and the structure of the supply chain have to be taken into account. Within the model CSR depends on four factors: diversity of the chain (product assortment, niche or mainstream market), claim to reach a certain level of performance standards, chain complexity, and power and influence of the company.

With the model, Cramer provides an approach to decide whether CSR measures should be implemented by the firm itself or in cooperation with other stakeholder groups. However, the instruments and several issues concerning the implementation are barely considered. Therefore, the usefulness of the step-by-step plan is limited, providing a comprehensive guideline for implementing CSR in the supply chain.

This contribution seeks to progress further than Cramer's model by providing an explorative analysis along different elements of the process of CSR realization. The status quo is studied in order to identify starting points to improve current CSR practices regarding their effectiveness. First, the main challenges for CSR on the national level of the supplier firms are considered. Triggers and motivation for CSR measures in the supply chain are elaborated as the second process element. Further, implemented CSR instruments, realization degree of CSR measures and implementation issues are pointed out. In this context, incentive mechanisms for CSR practices are examined as well. The governance instruments of the focal firm, such as control mechanisms and sanctions, are considered in the fourth analysis step. Suppliers are faced with social requirements by focal firms and have to comply. How suppliers respond to such requirements is the fifth process element. In this context it is analyzed whether the reaction of the supplier firms causes feedback and alterations to implementation practices of CSR in supplier firms.

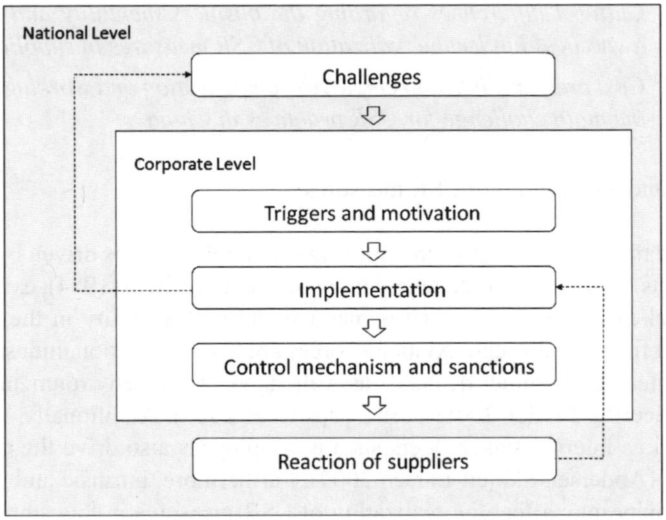

Fig. 2. Process elements of CSR realization in the supply chain

3 Propositions concerning CSR realization in supply chains

Before the presentation of the experts interviews results, some propositions are derived from scientific literature, which are proved through the analysis later on. In order to contain the research field, this contribution concentrates on social aspects of CSR measures in Chinese supplier firms.

3.1 CSR Challenges

In the first step, the nation-specific challenges for realizing CSR in China have to be examined. Lam (2009) identifies the following key challenges: Chinese-specific business behavior, intellectual property rights issues, internal due-process for dealing with ethical issues, lack of Chinese government support and missing knowledge of managers regarding CSR in Chinese supplier firms.

Chinese companies are even nowadays broadly perceived as low-cost manufacturers that hardly care about labor conditions (Lübcke et al. 2007). Park-Poaps and Rees (2009) mention three conditions which promote unfair employment practices in supplier firms: labor intensive and limited-automation production, competitive pressure to lower production costs, and complex production networks with multiple layers of subcontracting firms, which limits the control at the bottom level of the production chain.

Chinese supplier firms have to cope with a strong rivalry in the supply market. Furthermore, the business attitude is a challenge for realizing CSR in China. Short-term perspective and short-term goals are prevalent in Chinese firms. This business mentality affects the management and puts managers under pressure to fulfill these expectations (Lam 2009). Often, the positive consequences of CSR measures become visible only in the medium or long term while costs occur in the present. Therefore, the supplier has to cope with a conflict of objectives.

Proposition 1: Cultural differences regarding the business mentality and missing CSR awareness hinder the realization of CSR measures in supplier firms.

Proposition 2: Cost pressure linked to employee compensation and working conditions is the main challenge for CSR practices in China.

3.2 Triggers and motivation for CSR measures

In general, the motivation for CSR measures in the supply chain is driven by internal and external aspects. Following the stakeholder approach of Freeman (1984), external triggers are central stakeholder groups which demand social responsibility in the supplier networks of focal firms. In this context major forces are legal regulation, industry peer pressure, media interest, consumer demands as well as NGOs like environmental and social groups (e.g. Seuring/Müller 2008; Park-Poaps/Rees 2009). Additionally, investors and unions as well as internal stakeholders such as employees also drive the CSR commitment of firms (Andersen/Skøjett-Larsen 2009). Furthermore, intrinsic ambition and corporate philosophy may affect the realization of CSR measures in the supply chain (van Tulder et al. 2009).

Further triggers to implement CSR in the supply chain are gaining competitive advantage, improving firm performance and product quality (Walker et al. 2008) and limiting potential reputation losses (Seuring/Müller 2008). Firms promote CSR within their supplier network to increase control and minimize risks in order to guarantee the safety or quality of their products (Cramer 2008). Upstream suppliers are the determining factors. Lack of compliance with labor, environmental, health and safety standards may result in risks regarding security and sustainability and may also be reasons for reputation losses (Roberts 2003). Nowadays, the modern information and communication technologies facilitate the distribution of information across national borders (Schneider et al. 2008). Therefore, unethical business practices such as human rights abuses, unsafe working conditions, violation of union rights, child labor etc. are increasingly difficult to hide (Andersen/Skøjett-Larsen 2009). Welford and Frost (2006) identify risk reduction as the main driver for CSR measures along the supply chain. Western focal firms cannot afford to be perceived as firms which harm employees or environment in the supply chain. Bad publicity may have a negative impact on reputation and may damage brands (Millington 2008). Especially, business-to-consumer business models depend on brand and the image is elementary for success in the market.

Proposition 3: Risk management to protect image and reputation of firms is the main trigger for the engagement in CSR measures in the supply chain.

3.3 Realization of CSR

The way of realizing CSR in the supply chain depends on the diversity and complexity of the supplier network. Further, the power of influence by focal firms determines the implementation of CSR measures. However, there is a development from adversarial relationships between a firm and its suppliers toward stronger and closer relationships between one firm and fewer suppliers (Chen/Paulraj 2004). Focal firms as key purchasers have a greater importance and therefore a higher influence than one smaller purchaser among many others (Cramer 2008).

The principal supply chain decision is the selection of supplier firms. The main criteria in the supplier selection process are price, quality, delivery, and linked services (Beamon 1999). In recent years, labor practices of suppliers gain in importance (Yu 2008). In order to meet the requirements of the global market and several stakeholder groups firms do not only select their direct suppliers, they also have to consider the entire supply network (Hutchins/Sutherland 2008).

A supply chain consists of independent partners which carry out one or more processes for the production of the final product. In order to organize and coordinate the supplier network the focal firm depends on the information from the suppliers. In this context, focal firms are often faced with information asymmetry. Often information by suppliers is not available or incomplete. The partners in the supplier network could act opportunistically in terms of social and environmental requirements and may therefore constitute a risk for the focal firm's reputation (Ciliberti et al. 2009).

Codes of conduct are the most popular tool to realize CSR in the supply chain. They define required behaviors, responsibilities, and actions expected of suppliers and focus on ethical, social and environmental issues. Further instruments to ensure the compliance with social and environmental requirements within the supply chain are standards and management systems such as SA 8000 (social accountability and ethical working place condition), OHSAS (Occupational Health and Safety Assessment Series), EMAS (Eco Management and Audit Scheme) and ISO 14001 (Ciliberti et al. 2009).

Compared to quality and environmental management practices the management process of social issues is often fragmented and not standardized (Kortelainen 2008). Lee and Kim (2009) point out that environmental standards such as ISO 14001 are more often implemented and certified in Korean supply chains than social standards. They state that there is a lack of integration of social standards in supplier networks. Also, Zhu et al. (2005) register an increased environmental awareness in China. Additionally, Kortelainen (2008) points out that firms rarely have well-defined and documented management systems for social issues apart from occupational health and safety.

Focal firms often introduce codes of conduct in their supply chain networks. Suppliers have to comply with these codes and endure several audits as part of their contracts (Jiang 2009). Standards and CoC have many advantages and may improve working conditions and business practices. Nevertheless, codes are not a fast-selling item and sometimes may even have a counterproductive effect (Yu 2008, Ciliberti et al. 2009). Less is known about conditions under which standards and CoC lead to alteration of business practices and contribute to a more responsible behavior of different tiers of suppliers. Table 1 summarizes the most frequently mentioned advantages and disadvantages of standards and CoC.

Seldom, suppliers are integrated in the development process of such codes. This may lead to unachievable targets set by the focal firm and to relocation of the responsibility to the suppliers (Jiang 2009). The introduction of labor-related codes can result in a "race to ethical and legal minimum" labor standards and may at the same time affect the employees negatively (Yu 2008). The dialogue between buyers and suppliers raises awareness and reduces defensive demeanor of suppliers (Andersen/Skøjett-Larsen 2009). Van Tulder et al. (2009) suggest that involvement of stakeholders, in this case suppliers, in the design process of CoC increases the acceptance and probability of realization.

Tab. 1. Advantages and disadvantages of standards and CoC

Advantages	**Disadvantages**
– Definition of rules and procedures – Support of communication and coordination between partners – Reduction of information asymmetries and improvement of incompleteness of contracts – Decrease of transaction costs (searching, monitoring, bonding, enforcement costs) – Improvement of transparency – Benchmarking opportunities – Business process change – Improvement of working conditions	– Enforcing 'Western values' – External influence of buyers – Additional costs for supplier – Missing understanding of requirements by suppliers – Relocation of the cost in adverse manner for employees – Lack of inclusion of on-site situation in the respective country – No participation of suppliers in the development process – Control of compliance is still difficult

(see Egels-Zanden 2007; Helin/ Sandström 2008; Yu 2008; Ciliberti et al. 2009, Jiang 2009, Pertersen/Krings 2009; Van Tulder et al. 2009)

Proposition 4: Dialogue and experience exchange with suppliers lead to higher acceptance of CSR measures in the supply chain.

A lack of transparency may enhance the occurrence of opportunistic behavior of suppliers (Ciliberti et al. 2009). Therefore, most focal firms control compliance with stipulated codes in the form of audits (Jiang 2009). Auditing is a commonly used method to ensure compliance by setting criteria for quality, environmental and/or social issues, and to identify improvement opportunities (Kortelainen 2008).

However, it has been observed that Chinese suppliers frequently do not comply with codes. The reasons are manifold: some suppliers are simply not able to fulfill the requirements caused by linked costs and missing management knowhow. Other suppliers consciously breach the CoC (Jiang 2009).

According to Spence and Bourlakis (2009) supply chain responsibility has not yet been achieved in practice. Therefore, it is important to analyze whether the current instruments for realizing CSR in the supply chain are sufficient and how to improve the effectiveness of the measures.

4 Analysis of CSR realization in Chinese supplier firms

This paper aims to analyze the status quo realization as well as the effectiveness of CSR measures in the supply chain of multinational corporations (MNC). CSR measures in Chinese supplier firms are focused because of the importance of China as producing nation and the frequent integration of Chinese suppliers in global supply chains.

4.1 Methodology

In the context of the analysis the transnational model of CSR (Arthaud-Day, 2005) is used which integrates three key aspects. First, the strategic orientation is a multinational corporation approach and CSR is analyzed from a multinational perspective. The second aspect is the content domain. In this context, human rights, labor, and the environment are the three dominating areas which are focused on in the international CSR literature. And third the general perspective, MNC activities are evaluated on an ideological opinion, operational processes, or societal demands (Arthaud-Day, 2005).

In order to gather information regarding the implementation of CSR measures in Chinese supplier firms and to answer the research questions, interviews with eleven supply chain experts were conducted in 2009. The group of interviewed experts was diversified and included corporate representatives of large focal firms such as Adidas, Puma, Tchibo and Arcandor, representatives of NGOs, for example from SAI (Social Accounting Initiative), of industry initiatives like ICTI (International Council of Toy Industries) and CNTAC (China National Textile and Apparel Council), as well as managers from Chinese suppliers such as Esquel.

This multi-perspective view was chosen to obtain insights into the issue of CSR realization in Chinese supplier firms. The interviews were conducted in a semi-structured manner and took between 70 and 120 minutes each. In addition to general questions regarding CSR in China, essential topics of the interviews were:

1) Challenges for the realization of CSR in China,
2) Triggers and motivations for CSR in the supply chain, including the specific challenges in Chinese supplier firms,
3) Implemented CSR instruments in the supplier firms, including the types of implemented instruments and seemingly appropriate incentives to support the compliance with CoC and standards,
4) Control mechanisms for the compliance with CoC and the types of sanctions used by focal firms against suppliers in the case of non-compliance,
5) Reactions of suppliers, concerning the questions of whether Chinese suppliers consider CoC and standards as external interference with their own authority.

Therewith, the topics reflect the process elements of CSR realization in the supply chain which were already introduced in Part B of this paper (see fig. 2).

4.2 Findings from the interviews

4.2.1 Challenges for CSR in Chinese supplier firms

Working conditions continue to be one of the main problems in Chinese supplier firms, even if there have been considerable changes. From a CSR perspective, improvements and alterations are often realized in the so-called hardware part. Hardware means aspects such as health and safety manner or environmental concerns. Measures in these areas, for example the installation of fire extinguishers, are more visible, easier to realize, to control, and to evaluate. In some cases, the handling of resources and chemicals has improved through risk awareness trainings and legal rules of action. But still there are many firms that, for example, discharge their raw sewage into the rivers and go unpunished. Labor conditions are changing in the production plants as well. There are positive examples, for example in the textile industry, where working conditions equate more and more to those in Europe. In spite of the improvement in recent years, social issues like working conditions usually are a constant problem in Chinese supplier firms.

Overtime is frequent in China. As an issue, it has two dimensions. The first dimension includes requirements from the workers. In the context of overtime, the phenomenon of migrant workers has to be taken into account. Workers come from distant provinces and want to work in order to earn as much money as possible. Later on they move back to their home provinces. Among these migrant workers the tolerance for overtime is higher and even desired to a certain extent, as long as it does not exceed the sufferable quantity. On the other hand the overtime issue cannot entirely be blamed on the suppliers or their employees. In fact, the buying and order behavior of the focal firm is crucial for the extent of the overtime. Often focal firms give short-dated orders or order changes. There are also seasonal variations in the order volumes. Therefore, the responsibility lies not only with the supplier but also with the focal firm. The issue of overtime is influenced by three groups: workers, buyers (e.g. focal firms) and supplier firms. If only suppliers aim to reduce working hours there will be no success until the other two parties are changing their attitude as well. In general, if overtime does not exceed a certain extent, it is not a critical issue in China.

Appropriate wages and overtime compensation are major problems in China. The local governments set a minimum wage which can usually be met by the supplier firms. Often these minimum wages do not suffice for an appropriate livelihood. Further, overtime pay as required by law is quite high. Against the background of global competition and cost pressure these overtime payments are seldom affordable for supplier firms.

Another issue between focal firms and Chinese suppliers is mutual trust. Buyers use audits as assurance for compliance with environmental and social requirements. Even if suppliers build up own management systems, conduct own CSR measures and join domestic or international CSR initiatives, Chinese suppliers still have to pass audits. Self-monitoring efforts such as CSR reports are not accepted by buyers. Often suppliers have to pay for audits by themselves. In times of a worldwide financial crisis, multiple audits per year and the associated costs may lead to bankruptcies of Chinese supplier firms.

The Chinese business culture is another challenge for the realization of CSR in Chinese supplier firms. Often the managers of supplier firms pursue short-term goals. "Making money as soon as possible" is a prevalent business premise. In this context, the realization of middle and long-term CSR measures is difficult.

A further issue in China is that suppliers forward orders to subcontracting firms, which means that parts of the order volume are outsourced. Often the focal firm is not informed about such subcontracting and has no impact or control. Nothing is known about working conditions in these subcontracting firms.

In developing nations like China child labor is often perceived as a main issue. The interviewed CSR experts agreed that currently child labor in China is not an issue as it is more and more decreasing.

Law enforcement is another significant problem in China. The Chinese government passed a well elaborated labor law. The requirements are strict and unrealistic in some ways, for example, regarding working hours. The labor law stipulates maximum working time at 44 hours per week and 36 hours overtime per month. For suppliers it is difficult to comply with these requirements, especially if competitors do not comply or if the order volumes and delivery times of buyers are volatile.

Different interviewed experts identified social insurance for migrant workers and medical care as major problems in China, requiring a solution. The Chinese law requires every employee to have a social insurance account, payments for which are deducted as a certain percentage of the worker's salary, while the company has to contribute a certain percentage, too. This social security account is not transferable from province to province. Therefore, it affects the wages of the migrant workers negatively and they cannot benefit from it. In this context, it has to be noted that these are primary duties of the national government. It is questionable whether this should be a part of the corporate responsibility.

The findings regarding the main challenges of CSR are confirmed for the most parts by the current scientific literature. Welford and Frost (2006) point out that the lack of law enforcement is an ongoing problem. Therefore, requirements regarding working hours, minimum wages, overtime payments, health and safety are undermined by supplier firms. Lam (2009) highlights that continuous cost pressure and business culture in China are limiting factors for the realization of CSR practices. In this regard, the lack of awareness

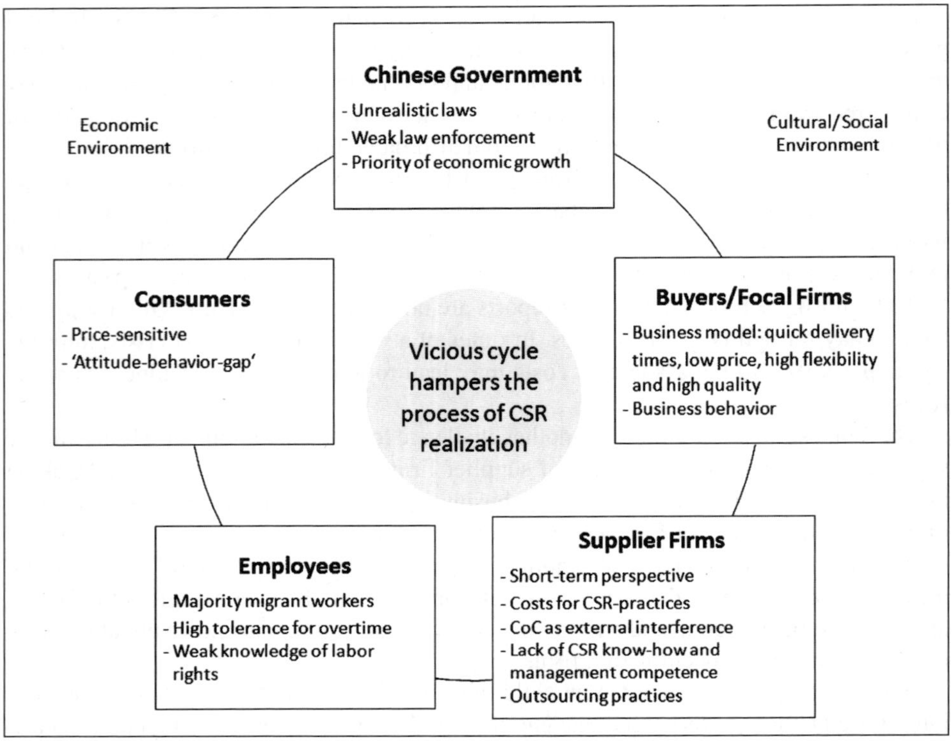

Fig. 3. Vicious circle of stakeholder influence and economic and cultural conditions

and knowledge about CSR benefits are heightening factors as well. Therefore, the first proposition that cultural differences and missing CSR awareness hinder the realization of CSR is supported by the finding from the expert interviews.

Production costs, delivery times and quality are still the main supplier selection criteria for focal firms. In the context of the financial crisis, order volumes of Western firms in China decreased. Moreover, the rivalry among Chinese supplier firms increased. The cost structure of a supplier firm is crucial in the light of competition. Due to the missing CSR education, measures regarding social issues are perceived as additional costs. Hence, the second proposition, namely cost pressure combined with employee compensation and working conditions being the main challenge for CSR practices in China, is supported by the interview results.

The complex of problems of medical care and social insurance is rarely mentioned in the scientific literature as being a CSR challenge. The reason is probably because these are government obligations. Then again, there have to be considerations how supplier firms can support their own employees in this situation.

On the basis of a stakeholder approach a vicious circle can be identified (see fig. 3) which presents a barrier for the CSR realization in China.

4.2.2 *Triggers and motivations*

The analysis of the qualitative interviews points out that the main driver for CSR in the supply chain is the goal of risk minimization to protect corporate reputation. Therefore, proposition three, which identifies risk management as the main trigger for CSR implementation is obviously supported by the statements of the supply chain experts. The protection of the corporate image has priority for focal firms. The efforts of focal firms to encourage suppliers to comply with social and environmental requirements mainly intend to prevent scandals and bad publicity.

In the opinion of the interviewed experts, pressure from NGOs and media as well as the corporate philosophy are crucial triggers to engage in CSR, too.

Tab. 2. Overview of triggers on CSR measures in supplier firms

Triggers	Percentage
Risk management	90%
Corporate philosophy	46%
Pressure from NGOs	46%
Competitive differentiation	36%
Pressure from media	27%
Consumer expectations	18%
Guidelines/legal requirements	9%
Investor expectations	0%
Peer pressure/competitors do it, too	0%

Corporate representatives specifically accentuated corporate philosophy as an important internal trigger for CSR measures. However, especially in family-owned small and medium-sized enterprises (SME) CSR practices are more often intrinsically motivated and CoC are a direct consequence of the ethical principles of the company owner (Ciliberti et al. 2009).

Consumer expectations are often mentioned as a driver for CSR practice. According to corporate CSR-experts consumer and investor expectations are rarely perceived as direct triggers or considered as success factors. This coincides with the findings of Devinney et al. (2006), who showed that if CSR is measured via opinions, a very positive picture emerges. But when CSR is measured using behavior, a significantly different picture appears. This phenomenon is also labeled "attitude-behavior-group". Also Welford and Frost (2006) refer that pressures from consumers are seldom noticeable for managers and therefore not such a decisive factor for CSR realization. Consumers and investors perceive missing CSR measures negatively, and it is more of a disappointment criterion (Bhattacharya/Sen 2004).

The Chinese government realized the potential long-term benefits of CSR programs and has been actively engaging in a campaign to foster CSR engagement of domestic firms and multinational enterprises. Focused areas of the government campaign are labor, buyer relations, tax, and environment (Zheng/Chen 2006). But until today, law enforcement and transparency are poorly developed in China (Kimber/Lipton 2005). Furthermore, the level of corruption remains high (Luo 2006). This assertion was affirmed by the interviewed members of NGOs. Therefore, legal requirements act less frequently as triggers for the implementation of CSR measures.

4.2.3 Implemented CSR instruments

CoC are the most frequently implemented instrument for CSR in the supply chain. The implementation of standards and management systems such as ISO 14001 and SA 8000 is still uncommon. Chinese supplier firms mostly focus on compliance with codes and on passing audits successfully. However, codes do not address all identified challenges in Chinese supplier firms. Furthermore, they are too narrow in certain dimensions for the current development stage of some Chinese supplier firms. Specifically regarding overtime and appropriate wages suppliers can seldom meet the requirements. The alteration of business processes taking into account codes of conduct and further social issues is linked to costs for suppliers. Therefore, many Chinese suppliers do not comply or do so only partially.

Interviewed corporate representatives suppose links between implemented CSR practices and quality of products due to the adaptation of processes and increased efficiency. For instance, extension of the working days and excessive overtime exhaust employees and lead to a higher error ratio. The limitation of working hours and overtime may minimize the error ratio and improve the quality. The lack of awareness regarding the benefits of CSR, the focus on implementation costs of measures and periodic audits are main barriers for CSR realization.

Sometimes suppliers comply with the codes without deep changes in their business practices and processes. Therefore, they cannot benefit from the implemented CSR measures. The additional costs are reallocated within the company. This behavior may lead to large savings in certain areas and affect the working conditions for employees negatively (Yu 2008).

In order to examine incentive mechanisms for CSR realization the interviewed experts were asked for their opinions. The question concerning appropriate incentive mechanisms resulted in different answers. Trainings to increase awareness for CSR were mentioned as appropriate means. Especially, key personnel, such as managers should be educated regarding CSR issues (Andersen/Skjøett-Larsen 2009). Further, technical assistance for the implementation of CSR measures and improving production processes regarding CSR aspects are considered helpful incentive mechanisms as well.

Some interviewed experts emphasized that initiatives to control labor issues are ineffective if they are based on one-way communication and top-down approaches to govern the issues (Jørgensen et al. 2003). Long-term relationship building in the sense of partnership may support mutual trust and respect as well as lead to fair and reasonable business practices. The partnership approach indicates strategic relationships between buyers and suppliers to share goals and benefits and to acknowledge mutual interdependence for successful CSR realization (Mohr/Spekman 1994). According to Poaps and Rees (2009) the partnership approach to support appropriate working conditions may affect bottom-level suppliers by influencing all levels of chain tiers.

Another means to increase compliance with social and environmental requirements may be financial incentives, for example compensation of realization costs through increased prices for products. Such an incentive mechanism is difficult to realize because price pressure is prevalent in supplier – buyer relationships. The increase of order volumes and the extension of the scope of products requested are further methods to set incentives for codes compliance. Andersen and Skjøett-Larsen (2009) perceive enlarged order volumes as an effective incentive mechanism, too.

If social and environmental criteria are a mandatory part of the supplier selection process and compliance with codes lead to a competitive advantage, this may provide incentives for suppliers to comply with requirements. The interviewed experts share the opinion that the prospect to become a strategic supplier stimulates the involvement with CSR practices. Long-term contracts are important for the survival of suppliers in the market and not least for the buyer firm (Carter/Roger 2008). If compliance with CoC is a precondition for these contracts, suppliers will increase their CSR engagement.

Cost sharing concerning audits is a double-edged sword. In this context the opinions of the interviewed experts diverge. Corporate representatives and industry initiatives consider the absorption of audit cost through suppliers as a way to support CSR engagement. Due to their own financial contribution, suppliers will be interested in passing the audit in the first run. Moreover, Western focal firms have often several thousands of suppliers. Hence, it is not possible to bear the audit costs for all suppliers. The NGO members as counterparts postulate that the focal firms should bear at least part of the audit costs.

Peer pressure among suppliers may increase the compliance of CoC as well. Moreover, cooperation with other buyers in order to increase the pressure on the suppliers can enforce the realization of codes as well. The BSCI (Business Social Compliance Initiative) is an example for such buyer cooperation.

4.2.4 Control mechanisms and sanctions

A further finding is that compliance with CoC is usually controlled via audits. This result is confirmed by the current scientific literature (e.g. Andersen/Skjøett-Larsen 2009; Jiang 2009; Spence and Bourlakis 2009). Normally, an initial audit is conducted. If the audit shows non-compliance with the corporate code, a corrective-action-plan will be developed to faster improvements. After some months re-audits are performed to evaluate the corrective measures taken.

An ideal engagement and auditing process is provided in fig.4. The different phases of the process are accompanied by a continuous dialog between the involved parties. Such intensive dialog takes time and demands a high level of involvement of all parties. Especially, for companies with a large number of suppliers it is impossible to execute such an intensive dialogue-oriented approach with each of them. Therefore, focal firms have to select carefully which suppliers could act as multiplicators of responsible practices into the overall network.

Chinese firms often supply different Western companies. Therefore, they have to comply with different CoC (partner-specific, industry-specific, cross-sectoral) at the same time and have to pass several audits per year. Often the suppliers have to bear all or at least the main portion of the audit cost by themselves.

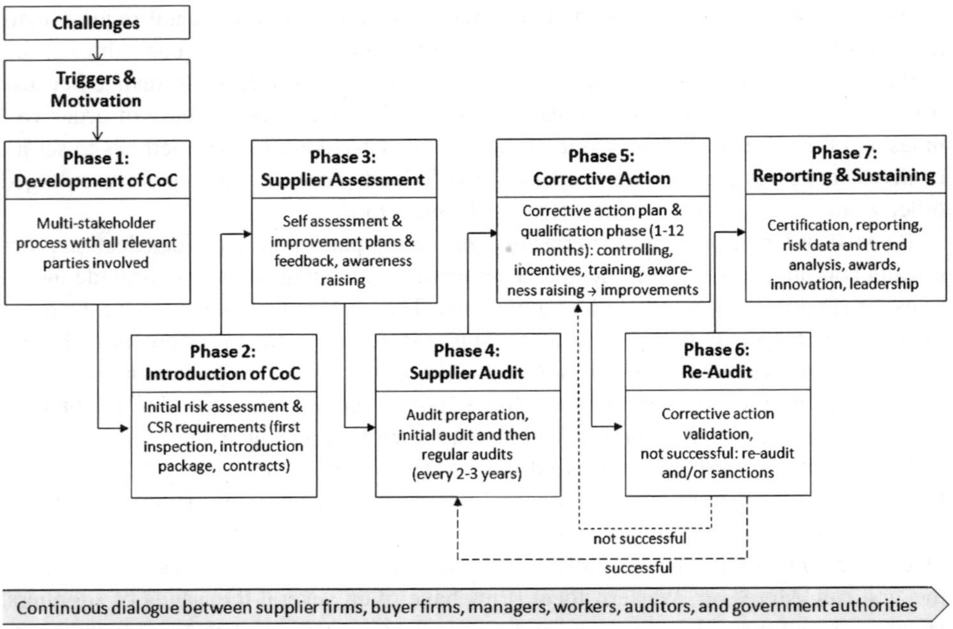

Fig. 4. Ideal engagement and auditing process

The most effective enforcement mechanism for the compliance with CoC would be the cancellation or at least temporary disruption of the business relation. This kind of sanction may increase the pressure on the suppliers. However, due to the mutual interdependence this is quite difficult to realize, because the focal firm itself has deadlines and delivery times to meet as well.

In fact in cases where Chinese suppliers ignore requirements of the focal firms, and thus do not comply, a temporary exclusion from the supply chain network may be the consequence. However, focal firms often support the stepwise improvement of working conditions. The improvement of business practices is a process and cannot be realized overnight. Therefore, the exclusion from the supply chain rarely happens. However, in case of noncompliance the reduction of order volume is a noticeable sanction mechanism.

4.2.5 Reactions of suppliers

Initially, most of the Chinese supplier firms consider CoC as external interference with their own authority and as additional costs. The short-term perspective is prevalent in Chinese firms. Benefits of CSR activities, however, can mostly be observed in the longer run. Therefore, Chinese suppliers perceive CSR as a competitive disadvantage. Costs and non-observable benefits are the main reasons for resistance and non-compliance. These reactions of the suppliers lead to a feedback effect concerning the implemented CSR measures.

Increasingly, there are initiatives by Western focal firms, federal organizations, and NGOs to engage supplier firms in a dialogue. One example is the "Worldwide Enhancement for Social Quality (WE)" which is a public-private partnership (PPP) between the German Federal Ministry for Economic Cooperation and Development (BMZ), the German Technical Cooperation (GTZ) and the Tchibo GmbH (one of Germany's biggest international consumer goods and retail companies). The overarching goal of the WE-project is the improvement of working conditions in Asian manufacturing sites by following a training approach which emphasizes a dialogue between all parties as the key success factor.

Trainings and experience exchange increase the understanding of CSR benefits and create awareness. Through guidance and education about CSR instruments and CSR effects the resistance of suppliers is declining. Further, integration of CSR aspects into internal processes leads to the revision and alteration of conventional processes and may increase efficiency. Support and guidance may activate suppliers to introduce proactive and self-developed CSR-measures. There seems to be a shift from a purely standard-based approach to more dialogue orientation in order to create awareness and to support the compliance with CoC. These findings from the interviews analyzed confirm the fourth proposition, namely that dialogue and experience exchange with suppliers lead to higher acceptance of CSR measures in the supply chain.

The process of reaching a higher level of CSR awareness and a higher degree of CSR realization is demonstrated in fig. 5. The decisive factors are dialogue and trainings complemented with an internal process change (e.g. optimization of the production process within the supplier firms and realistic delivery and price demands in the buyer firm) in both the supplier and the Western buyer firm. As a consequence, the compliance with CoC will gradually increase.

The development from a standard-based to a dialogue-orientated CSR approach is also in line with results of other researchers (e.g. Andersen/Skjøett-Larsen 2009; Poaps/Rees 2009). The likelihood of appropriate and proactive CSR measures in Chinese supplier firms is linked with the CSR education and knowledge of the respective managers (Lam 2009).

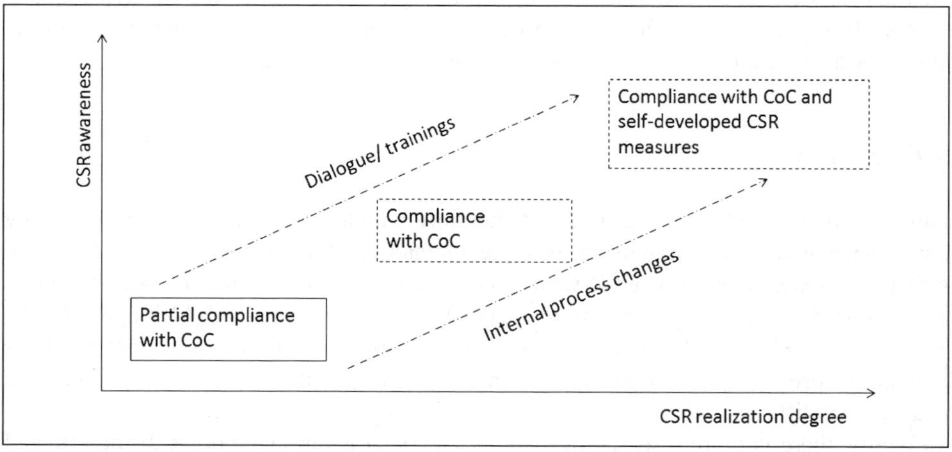

Fig. 5. From a standard-based CSR approach to dialogue-orientation

In this context it has to be mentioned that a dialogue-based approach may be risky. The processes have to encompass the internal business process of both the supplier and the focal firm. Ever since the example of Nike´s early attempts to introduce CoC in its supplier network it is known that the business model of the focal firm is the decisive factor for the implementation of CSR. In the case of Nike it was not possible for the supplier to comply with the CoC because of their disconnection with Nike's performance incentives which based solely on price, quality and delivery times (Zadek 2004).

Furthermore, it is probably too costly and too complex for a buyer to realize a dialogue-oriented approach with all suppliers of its entire supply chain. For example, IKEA has about 1,500 suppliers in 55 countries (Andersen/Skjoett-Larson 2009), Tchibo has about 950 first-tier suppliers around the world, and Otto has even 2.500 to 3.000 first-tier suppliers. Obviously, a dialogue-oriented approach with each supplier is impossible to manage. Hence, it will be more appropriate to limit a dialogue-oriented approach to a manageable number of supplier firms. The buyer firms may choose the corresponding firms due to their strategic significance and their potential capacity to pass the experiences on to other (second- and third-tier) suppliers within the network.

The interviewed corporate representatives have supported this assumption by stating that they will reduce the number of their suppliers within the next few years in order to shift from an arm's length to a true partnership approach, in which a supply chain partnership is placed on a more strategic level.

Thus, Western companies should reconsider their approach of having a uniform CoC regardless of the specific local conditions in the country they operate in. More realistic is a procedure of incremental improvement targets accompanied by intensive training measures to raise the management competence on the one side and the awareness for the advantages of CSR practices on the other. Therefore, corrective action plans combined with the respective training measures and the adaptation to the specific CSR level of the supplier might be a more promising way of CSR realization than a pure standard- and control-based approach.

In summary, the described way of a more dialogue-oriented approach which includes all important stakeholders combined with training measures and step-by-step requirements adjusted to the specific situation will have the potential to break the vicious circle and generate a virtuous circle of responsible supply chain management.

5 Conclusion

Although there is an emerging body of literature which focuses on responsible supply chain management, a lot remains to be done. Greater emphasis needs to be placed on the effective implementation of CSR within the supply firms rather than on the normative ethical justification of CSR in the supply chain and the content of CoC (Millington 2008). There is a need for research in countries where training and experience of CSR practices may not be present and managerial problems were caused due to substantial cultural and institutional differences between the leading buying company and the supplier firms. Additionally, there is more research needed which considers the viewpoint of the affected firms in the supply chain network abroad.

This paper makes a considerable contribution by: 1) providing a comprehensive literature review regarding the implementation of CSR within the supply chain; 2) analyzing the process elements of CSR realization in a multi-perspective view considering not only the buyers but different participating parties; 3) considering sanction and incentive mechanism as important elements within the overall supply chain management process; 4) pointing out the great importance of awareness raising for CSR on the supplier side and dialogue between all participating parties before and during the implementation process.

In this paper the realization of CSR measures in Chinese supplier firms is analyzed through expert interviews. An overview of the process elements of CSR implementation has been provided. The most frequently used CSR instruments are codes of conduct. The findings suggest that codes of conduct broach only some of the major nation-specific challenges in China such as working conditions, wages and overtime. Even if these aspects are covered by the applied codes the enforcement of CoC is a prevalent problem. Moreover, CoC can have counterproductive effects on CSR realization. Suppliers react with resistance and non-compliance or only meet some aspects of the codes of conduct while ignoring others. Focal firms put the responsibility for non-compliance on the supplier firms. They apply their codes to all suppliers in their supply chain regardless of differences concerning economical development and country-specific conditions. However, the business behavior of focal firms determines the compliance as well. High cost pressure, short-dated orders and short-term variations of order volumes hinder suppliers to comply with CoC. Furthermore, suppliers have to bear the audit costs by themselves. The lack of CSR awareness and the understanding of linked benefits increase the resistance of suppliers regarding CoC.

The common purely standard-based approach seems less successful than a more dialogue-based approach. Training and experience exchange between suppliers is appropriate to create awareness for CSR and to increase the understanding of CSR benefits. Technical assistance for the implementation of CSR measures and the alteration of corporate processes may lead to more efficiency and better quality as well as support suppliers to benefit more from CSR measures. The integration of suppliers in the code's development process may improve relationship building and mutual trust as well as the compliance with the code. Furthermore, the code may then better fit the prevalent challenges in the respective supplier's country. In order to reduce audit costs multiple audits should be avoided. Therefore, cooperations with other buyers may lead to a reduction of audit frequency and linked costs for suppliers.

The generalization of the findings may be limited due to the number of conducted interviews. On the other hand the interviews with supply chain experts provide a deep insight into the challenges and processes of CSR realization in Chinese supplier firms. The limitation regarding generalization may be reduced through further interviews with experts in the future. Nevertheless, this explorative study is a starting point for future research and the development for appropriate instruments to increase the CSR realization in supplier firms in global supply chains. The results give managers of buyer firms a better understanding of the overall supply chain management process, the interrelation of the process elements, and the significance of measures that support or hinder the compliance with CoC and hence the realization of CSR. A supplier firm has to participate in the formulation and implementation of CoC, mainly to enhance its own market position and its value for current and potential business partners.

References

Andersen M, Skjøett-Larsen T (2009) "Corporate social responsibility in global supply chains". Supply Chain Management: An International Journal 14(2):75–86

Arthaud-Day M (2005) "Transnational Corporate Social Responsibility: A Tri-Dimensional Approach to International CSR Research". Business Ethic Quarterly 15(1):1–22

Bhattacharya CB, Sen S (2004) "Doing better at doing good: when, why, and how consumers respond to corporate social initiatives". California Management Review 47(1):9–24

Beamon BM (1999) "Measuring supply chain performance". International Journal of Operations and Production Management 19(3):275–92

Baughn C, Bodie N, McIntosh J (2007) "Corporate Social and Environmental Responsibility in Asian Countries and Other Geographical Regions". Corporate Social Responsibility and Environmental Management 14 (4):189–205

Bondy K, Matten D, Moon J (2008) "Multinational Corporation Codes of Conduct: Governance Tools for Corporate Social Responsibility?". Corporate Governance 16(4):294–311

Carter C, Jennings M (2002) "Social responsibility and supply chain relationships", Transportation Research, Part E 38:37–52

Carter CR, Rogers DS (2006) "A framework of sustainable supply chain management: moving toward new theory". International Journal of Physical Distribution & Logistics Management 38(5):360–387

Chen IJ, Paulraj A (2004) "Towards a theory of supply chain management: the constructs and measurements". Journal of Operations Management 22:119–50

Ciliberti F, de Groot G, de Haan J, Pontrandolfo P (2009) "Codes to coordinate supply chains: SMEs' experiences with SA8000". Supply Chain Management: An International Journal 14(2):117–127

Cramer JM (2008) "Organising corporate social responsibility in international product chains". Journal of Cleaner Production 16(3):395–400

Cruz J, Wakolbinger T (2008) "Multiperiod effects of corporate social responsibility on supply chain networks, transaction costs, emissions, and risk". International Journal of Production Economies 116:61–74

Davies IA, Crane A (2003) "Ethical decision making in fair trade companies". Journal of Business Ethics 45(1–2):79–92

Devinney TM, Auger P, Eckhardt G, Birtchnell T (2006) "The other CSR". Stanford Social Innovation Review, Fall, pp 30–37

Egels-Zandén N (2007) "Suppliers Compliance with MNCs Codes of Conduct: Behind the Scenes at Chinese Toy Suppliers". Journal of Business Ethics 75:45–62

Egels-Zandén N, Wahlqvist E (2007) "Post-Partnership Strategies for Defining Corporate Responsibility: The Business Social Compliance Initiative". Journal of Business Ethics 70:175–189

Freeman RE (1984) "Strategic Management: A Stakeholder Approach". Boston: Pitman/Ballinger

Frenkel SJ, Kim S (2004) "Corporate Codes of Labour Practice and Employment Relations in Sports Shoe Contractor Factories in South Korea". Asia Pacific Journal of Human Resources 42(1):6–31

Jiang B (2009) "Implementation Supplier Codes of Conduct in Global Supply Chains: Process Explanations from Theoretic and Empirical Perspectives". Journal of Business Ethics 85:77–92

Jørgensen HB, Pruzan-Jørgensen PM, Jungk M, Cramer A (2003) "Strengthening Implementation of Corporate Social Responsibility in Global Supply Chains". The World Bank, Washington DC

Helin S, Sandström J (2008) "Codes, Ethics and Cross-Cultural Differences: Stories from the Implementation of a Corporate Code of Ethics in a MNC Subsidiary". Journal of Business Ethics 82:281–291

Hutchins MJ, Sutherland JW (2008) "An exploration of measures of social sustainability and their application to supply chain decisions". Journal of Cleaner Production 16:1688–1698

Kimber D, Lipton P (2005) "Corporate Governance and Business Ethics in the Asia-Pacific Region". Business & Society 44(2):178–210

Kortelainen K (2008) "Global Supply Chains and Social Requirements: Case Studies of Labour Condition Auditing in the People's Republic of China". Business Strategy and the Environment 17:431–443

Kovács G (2008) "Corporate environmental responsibility in the supply chain". Journal of Cleaner Production 16(15):1571–1578

Lam ML-L (2009) "Beyond Credibility of Doing Business in China: Strategies for Improving Corporate Citizenship of Foreign Multinational Enterprises in China". Journal of Business Ethics 87:137–146

Lee K-H, Kim J-W (2009) "Current status of CSR in the realm of supply management: the case of the Korean electronics industry". Supply Chain Management: An International Journal 14(2):138–148

Lübcke E, Ruth K, Yim I (2007) "Corporate Social Responsibility 'Made in China'". ITB Working Paper No. 60, University Bremen
Luo Y (2006) "Political behaviour, social responsibility, and perceived corruption". Journal of International Business Studies 37:747–766
Millington A (2008) "Responsibility in the supply chain", Crane A, McWilliams A, Moon J, Siegel DS (eds), The Oxford Handbook of Corporate Responsibility, Oxford University Press, New York, pp 363–383
Mohr J, Spekman R (1994) "Characteristics of Partnership Success: Partnership Attributes, Communication Behavior, and Conflict Resolution Techniques". Strategic Management Journal 15(2):135–152
Park-Poaps H, Rees K (2009) "Stakeholder Forces of Socially Responsible Supply Chain Management Orientation". Journal of Business Ethics, prepublication online 4th July 2009
Pertersen L-E, Krings F (2009) "Are Ethical Codes of Conduct Toothless Tigers for Dealing with Employment Discrimination?". Journal of Business Ethics 85:501–514
Prieto-Carrón M (2008) "Women Workers, Industrialization, Global Supply Chains and Corporate Codes of Conduct". Journal of Business Ethics 83:5–17
Roberts S (2003) "Supply Chain Specific? Understanding the Patchy Success of Ethical Sourcing Initiatives". Journal of Business Ethics 44:159–170
Sarkis J (2003) "A strategic decision framework for green supply chain management". Journal of Cleaner Production 11:397–409
Schneider A-M, Stieglitz S, Lattemann C (2008) "Social Software as an Instrument of CSR", in: Vaccaro A, Horta H, Madsen P (Eds) Transparency, Information and Communication Technology: Social Responsibility and Accountability in Business and Educations, Philosophy Documentation Center, Charlottesville, Virginia, pp 139–150
Seuring S, Müller M (2008) "From a literature review to a conceptional framework for sustainable supply chain management". Journal of Cleaner Production 16(15):1699–1710
Spence L, Bourlakis M (2009) "The evolution from corporate social responsibility to supply chain responsibility: the case of Waitrose". Supply Chain Management: An International Journal 14(4):291–302
Van Tulder R, van Wijk J, Kolk A (2009) "From Chain Liability to Chain Responsibility". Journal of Business Ethics 85:399–412
Waddock S, Bodwell CB, Graves SS (2002) "Responsibility: The New Business Imperative". Academy of Management Executive 16(2):132–148
Walker H, Di Sistob L, McBai D (2008) "Drivers and barriers to environmental supply chain management practices: Lessons from the public and private sectors". Journal of Purchasing & Supply Management 14:69–85
Welford R, Frost S (2006) "Corporate Social Responsibility in Asian Supply Chains". Corporate Social Responsibility and Environmental Management 13:166–176
Yu X (2008) "Impacts of Corporate Code of Conduct on Labor Standards: A Case Study of Reebok's Athletic Footwear Supplier Factory in China". Journal of Business Ethics 81:513–529
Zadek S (2004) "The Path to Corporate Responsibility". Harvard Business Review 82(12):125–132
Zheng Y, Chen M (2006) "China Moves to Enhance Corporate Social Responsibility in Multinational Companies", Briefing Series 11, China Policy Institute, University of Nottingham
Zhu Q, Sarkis J, Geng Y (2005) "Green supply chain management in China: pressures, practices and performance". International Journal of Operations & Product Management 25(5):449–468
Zutshi A, Creed A, Sohal A (2009) "Child labour and supply chain: profitability or (mis)management". European Business Review 21(1):42–63

ZfB
ZEITSCHRIFT FÜR
BETRIEBSWIRTSCHAFT

Fair Trade in Europe: The Role of Economics and Culture

Bert Scholtens

Abstract: This paper explores the proliferation of fair trade in Europe. We associate economic and cultural indicators to the extent of fair trade. We find preliminary evidence that it does not appear to be predominantly economic factors but rather socio-cultural factors that are to be associated with the development of fair trade in Europe.

Keywords: Fair trade · cultural indicators · economic development · Europe

JEL Classification: F02 · L31 · M14 · Z13

Prof. Bert Scholtens, Ph.D. (✉)
Department of Finance, Faculty of Economics and Business, University of Groningen, P.O. Box 800, NL-9700 AV Groningen, Netherlands; E-mail: l.j.r.scholtens@rug.nl.

1 Introduction

Fair trade is a global phenomenon. We try to identify determinants of fair trade at the country level and focus on Europe. With fair trade, the consumer includes characteristics about the production process in their spending decisions (Moore, 2004). Typical examples are artisan and crafts, and coffee, tea, bananas (see Littrell and Dickson, 1999; De Pelsmacker et al., 2005). Paying attention to 'non pecuniary' aspects of products is not new or limited to fair trade. Baker and Nofsinger (2002) investigate how social factors impact upon investment decisions. Becker-Olsen et al. (2006) find that affinity with environmental and sustainability issues affects environment behavior of households. Stern (2000), among others, argues how economic and social factors might impact on people's values and choices. However, there is little attention for the quantitative role of such variables at the macrolevel. In this note, we aim at determining the drivers of fair trade at the national level and associate the size of fair trade with economic and cultural indicators. First, we go into the definition of the concept of fair trade and its main characteristics. Second, we assess the extent of fair trade in sixteen European countries. Then we derive key indicators of fair trade and connect these with political-economic and cultural variables.

2 Fair trade

Fair trade in a microeconomic context suggests that protectionism and other trade distorting measures like import tariffs and barriers can be used to achieve a level-playing field (Bhagwati, 1996; Stiglitz, 2002). In some cases, the demand for compliance with environmental and labor standards can be motivated by a protectionist background where protectionists masquerade as moralists (Howse and Trebilcock, 1996). In contrast, we deal with fair trade as a way of conduct by producers, distributors and consumers who all engage in 'pro-poor' trade with predominantly developing countries (Littrell and Dickson, 1999). We regard as fair trade products those products that follow some criteria for social and environmental sustainability, established by the movement of fair trade importers and retailers (Moore, 2004). Here, fair trade tries to promote trade with producers who produce their products under certain conditions in developing countries. Consumers offer to pay higher prices as some kind of moral obligation to these producers. As such, fair trade is a moral obligation, which implies that there is no trade-off possible between these principles and consumer preferences (Lancaster, 1966; Jensen and Meckling, 1994). A typical feature of fair trade is the special role of the middlemen, the fair trade organizations (for an overview, see Leclair, 2002). These organizations also engage in labelling and branding of fair trade products. Bjørner et al. (2004) find that such labelling can be successful.

Fair trade organizations incidentally report about their activities. Fairtrade Labeling Organizations (FLO), International Fair Trade Association (IFAT), Network of European World Shops (NEWS!) and European Fair Trade Association (EFTA) give a combined overview of their activities (Krier, 2005). However, definitions differ and data are gathered differently. Therefore, the combination and comparison of the data has to be interpreted very carefully. In table 1, we report the basic statistics about fair trade in 16 European countries.

Tab. 1. Fair Trade in Europe

	Austria	Bel-gium	Den-mark	Germany	Finland	France	Ireland	Italy	Luxem-bourg	Nether-lands	Norway	Por-tugal	Spain	Sweden	Swit-zerland	UK	Total
Points of sale																	
Shops	100	295	6	800	25	165	6	500	6	412		9	95	35	300	100	2854
Supermarkets	2000	700	1000	23000	3000	10000	250	4000	80	3100	4000				2500	3100	56730
Others	1250	50	31	9800		132				1810		11	850	300	20	5000	19254
Total	3350	1050	1040	33600	3050	10300	260	4500	90	5320	4000	20	1090	340	2820	8200	79030
Paid staff (in fte)	48.5	83.2	5.3	203.3	32	85	9	128.5	2.4	106.7	3.5	0	29	17	63.1	245.8	1062
Turnover (in 1,000 €)	36.1	44.8	13.1	133	8.8	87.8	6.3	61.2	2.8	85.9	4.5	0.4	20.2	8.4	153.7	276.3	943
Market shares, in %																	
Coffee	2.3	1.7	2	1	0.4		2				1			<1.0	6	20	
Tea	2.1		<1.0	2			0.5				0.3			<1.0	5	0.5	
Bananas	2.3	4	0.9		5		0.5				1.5			<1.0	47	0.5	

Table 1 shows that there were 2850 worldshops in the 16 countries. Most of them were established in Germany, namely 800. Italy, the Netherlands, Switzerland and Belgium harbor about 300 to 500 fair trade shops. All other countries have less than 150 shops. An alternative indicator of the opportunities of customers to buy fair products is to investigate the number of points-of-sale. As such, it appears that the highest 'fair trade density' is in Switzerland, where there is one fair trade point-of-sale on every 48 thousand inhabitants. At the other end of the scale is Spain, where there are 2,1 million inhabitants per point-of-sale. A third main indicator is the turnover of fair products in relation to Gross Domestic Product (GDP). Here, turnover is highest in Switzerland before the Netherlands, the UK, Belgium, and Austria.

From Table 1, we conclude that fair trade constitutes only a very small part of the daily economy. In contrast to the claim by MacGillivray (2000), it has not penetrated substantially in any European economy. The highest penetration can be found in Central- and Northwestern European countries. Southern-Europe closes the ranks. These differences result from differences in market share of specific products. For example, fair trade bananas have a market share of almost 50% in Switzerland. Switzerland, not being a member of the EU and not subject to EU-policies with respect to the import of bananas, plays a crucial role in this respect. The market share of fair trade tea and coffee also is substantial in Switzerland, namely 5 and 6% respectively. Elsewhere, most market shares of the specific fair trade products are below 2%. An exception is fair trade coffee in the UK and the Netherlands (market share of 20 and 28% respectively), and fair trade bananas in Belgium and Finland (market share of 4 and 5% respectively). But the information about the market share of fair trade products can not always be regarded as reliable and can not be compared very well among the countries.

3 Drivers of fair trade

In this section, we assess where the apparent international differences in the 'taste' for fair trade might stem from. We first go into economic factors and then investigate cultural indicators. Our methodology is very simple. Given the limited quantity and quality of our datamaterial, we only come up with straightforward correlations. Of course, on this basis, we are not able to conclude about causal relations. This is a matter for future research.

3.1 Economic factors

We look into the following economic variables: gross savings and investments, these are proxies for the long term horizon that prevails in the economy; percentage of GDP spend on health as well as that spend on the military and military equipments, these can regarded as proxies for the type of preferences society hold; imports and exports in relation to GDP as a measure of economic openness; size of GDP and size of the population as a measure of size of the country; tax revenues as an indicator of the role of government in the economy; per capita carbondioxide emissions as an indicator of environmental quality in the country; per capita GDP as a measure of wealth; income share of the poorest 10% of inhabitants as an indicator of income distribution.

Tab. 2. Correlation coefficients political-economic indicators and fair trade indicators

	1	2	3	4	5	6	7	8	9	10	11	12	13	14	15	16	17	18	19	20
number of shops	-0.26	0.04	0.61	-0.18	0.64	-0.48	-0.09	-0.22	0.64	-0.35	0.09	-0.18	1.00							
points-of-sale	-0.39	-0.21	0.53	-0.28	0.75	-0.42	0.02	-0.18	0.79	-0.02	0.19	-0.26	0.77	1.00						
turnover	-0.31	-0.21	0.32	-0.28	0.60	-0.17	-0.15	-0.17	0.67	-0.38	0.42	-0.31	0.41	0.48	1.00					
staff	-0.43	-0.43	0.29	-0.28	0.80	-0.14	-0.07	-0.25	0.84	-0.35	0.43	-0.32	0.66	0.68	0.88	1.00				
inhabitants per worldshop	0.28	-0.41	0.19	-0.09	-0.13	-0.01	-0.27	-0.35	-0.19	-0.46	0.26	-0.18	-0.23	-0.18	-0.23	-0.25	1.00			
inhabitants per point-of-sale	0.27	-0.40	0.20	-0.11	-0.11	-0.01	-0.27	-0.35	-0.18	-0.30	0.27	-0.20	-0.22	-0.16	-0.23	-0.26	1.00	1.00		
turnover /GDP	0.09	0.51	0.33	0.13	-0.16	-0.37	-0.13	0.10	-0.11	-0.16	-0.23	0.12	0.19	-0.05	0.51	0.19	-0.24	-0.26	1.00	
inhabitants as a ratio of staff	0.27	-0.40	0.20	-0.11	-0.11	-0.01	-0.27	-0.35	-0.18	-0.30	0.27	-0.20	-0.22	-0.16	-0.23	-0.26	1.00	1.00	-0.26	1.00

Legend:

1 Gross fixed capital formation (% GDP)
2 Gross savings (%GDP)
3 Health expenditure (%GDP)
4 Imports of goods and services (%GDP)
5 Population
6 Tax revenues (%GDP)
7 CO_2 emissions (metric tons per capita)
8 GDP per capita, PPP (constant 2000 international US$)
9 GDP (current US$)
10 Income share of poorest 10% (% of total income)
11 Military expenditures (%GDP)
12 Export of goods and services (%GDP)
13 World shops
14 Total fair trade points-of-sale
15 Total turnover (€ million)
16 Total staff (fte)
17 Inhabitants per worldshop
18 Inhabitants per fair trade point of sale
19 Turnover fair trade shops/GDP
20 Inhabitants as a ratio of staff World shops

Table 2 presents the Pearson correlation coefficients of the economic indicators and basic fair trade measures. These measures either were discussed above or are combinations of these and other reference variables. The first four measures (13-16 in table 2: the number of fair trade shops, the number of selling points, the turnover in euros and the size of staff in fte) all are absolute numbers. The last four measures (17-20 in table 2: the number of inhabitants per fair trade shop, the number of inhabitants per fair trade selling point, the turnover of fair trade in relation to GDP, the number of inhabitants in relation to the size of fair trade staff) are relative indicators. Table 2 shows that the size of the country, as measured by population and gross domestic product, is strongly associated with the first four fair trade measures. This is not an unexpected result. However, surprising might be that almost none of the other political-economic indicators appears to be associated with these fair trade measures as most of them are slightly negative and weakly correlated among one each other. An exception is the expenditure for health as a percentage of GDP. This percentage is strongly and positively correlated with the fair trade measures. Furthermore, we find that the relative measures of fair trade appear to be correlated weakly and negatively in most cases. For the number of inhabitants per fair trade shop as well as the number of inhabitants per selling point, the negative sign is to be expected. However, we also find a negative correlation for the relative turnover and employment indicators. In all, it appears that macroeconomic indicators like wealth, government size, and openness are not associated with the (relative) size of fair trade in a country.

3.2 Cultural Factors

Data for cultural variables are derived from Hofstede (1980, 1991). His work consists of survey data about the values of people working in local subsidiaries of IBM in more than 50 countries. The actual surveys used in Hofstede (1980) date back to the 1970s. Updates and extensions have re-affirmed its main conclusions (see Hofstede, 1991). These data are used a lot in social and economic research (e.g. Garretsen et al., 2004). The fact that the data are more than 30 years old is not a main concern under the assumption that culture changes very slowly over time. Another reason to use these data is that they pertain to general features of culture for the countries in the sample. This suits our research objective since we want to emphasize the role of cultural values that are general and not specific to certain markets or transactions. Hofstede (1980) defines the following cultural indicators: *Power distance* measures social inequality and is the extent to which the less powerful members of institutions and organizations within a country expect and accept that power is distributed unequally. *Individualism* pertains to societies in which the ties between individuals are loose: everyone is expected to look after himself. *Masculinity* reflects the desirability for assertive behavior against the desirability of modest behavior. *Uncertainty avoidance* is the extent to which the members of a culture feel threatened by uncertain or unknown situations.

Table 3 shows the correlations of cultural indicators with eight key indicators of fair trade. There is a positive correlation between the cultural indicators and most fair trade variables, especially of individualism and masculinity. As to the relative measures, we establish that there no longer is a substantial positive correlation with masculinity. However, individualism still is significantly positive correlated with fair trade measures. Table

Tab. 3. Correlation coefficients cultural indicators and fair trade indicators

	1	2	3	4	5	6	7	8	9	10	11	12
number of shops	0.21	0.16	0.40	0.38	1.00							
points-of-sale	0.02	0.05	0.38	0.15	0.77	1.00						
turnover	0.09	0.61	0.49	-0.08	0.41	0.48	1.00					
staff	0.17	0.54	0.62	0.11	0.66	0.68	0.88	1.00				
inhabitants per worldshop	0.35	-0.35	-0.09	-0.02	-0.23	-0.18	-0.23	-0.25	1.00			
inhabitants per point-of-sale	0.31	-0.56	-0.19	0.04	-0.22	-0.16	-0.23	-0.26	1.00	1.00		
turnover /GDP	-0.11	0.41	0.33	0.10	0.19	-0.05	0.51	0.19	-0.24	-0.26	1.00	
inhabitants as a ratio of staff	0.31	-0.56	-0.19	0.04	-0.22	-0.16	-0.23	-0.26	1.00	1.00	-0.26	1.00

Legend:

1 Power distance
2 Individualism
3 Masculinity
4 Uncertainty avoidance
5 Number of World shops
6 Total fair trade points-of-sale
7 Total fair trade turnover (€ million)
8 Total staff
9 Inhabitants per worldshop
10 Inhabitants per fair trade Point-of-sale
11 Turnover fair trade shops/GDP
12 Inhabitants as a ratio of staff with World shops

3 suggests that fair trade prevails in societies that score high on individualism. This might suggest that it is relatively popular in societies where individuals take responsibility for what is going on in the world and try to have an effect by their behavior as a consumer. Although the correlation coefficients are almost all significant, there is, of course, no simple one-to-one relationship between fair trade and cultural indicators.

4 Conclusion

The fraction of fair trade products in relation to the overall consumption pattern is still marginal. At the national European level, we establish that it is not so much economic characteristics that are to be associated with the market share of fair trade, but merely cultural factors. This suggests that apart from access of developing countries' products to the world market, proper marketing is a key issue for the success of fair trade products. Explicitly linking the different non-pecuniary attributes of fair trade products to the cultural characteristics of consumers in the different countries might increase the use of fair trade products.

References

Baker HK, Nofsinger JR (2002) Psychological biases of investors. Financial Services Review 11:97–106

Becker-Olsen KL, Cudmore BA, Hill RP (2006) The impact of perceived corporate social responsibility on consumer behavior. Journal of Business Research 59:46–53

Bhagwati J (1996) Fair trade and harmonization: Prerequisites for free trade? Vol. I: Economic analysis: Introduction. MIT Press, Cambridge and London

Bjørner TB, Hansen LG, Russell CS (2004) Environmental labeling and consumers' choice – An empirical analysis of the effect of the Nordic Swan. Journal of Environmental Economics and Management 47:411–434

Garretsen H, Lensink R, Sterken E (2004) Growth, financial development, societal norms and legal institutions. Journal of International Financial Markets, Institutions, and Money 14:165–183

Hofstede G (1980) Culture's consequences: International differences in work-related values. Sage Publications, Beverly Hills

Hofstede G (1991) Culture and organizations: Intercultural cooperation and its importance for survival – Software of the mind. McGraw-Hill, London

Howse RL, Trebilcock MJ (1996) The fair trade-free trade debate: Trade, environment, and labour rights. International Review of Law and Economics 16:61–79

Jensen MV, Meckling WH (1994) The nature of man. Journal of Corporate Finance, 7:4–19

Krier J (2005) Fair trade in Europe 2005. FLO/IFAT/NEWS!/EFTA, Brussels

Lancaster KJ (1966) A new approach to consumer theory. Journal of Political Economy 74:132–157

Leclair MS (2002) Fighting the tide: Alternative trade organizations in the era of free global trade. World Development 30:949–958

Littrell MA, Dickson MA (1999) Social responsibility in the global market. Fair trade of cultural products. Sage Publications, Thousand Oaks

MacGillivray A (2000) The fair share: The growing market share of green and ethical products. New Economics Foundation, London

Moore G (2004) The Fair Trade movement: parameters, issues and future research. Journal of Business Ethics 53:73–86

Pelsmacker P, Driesen L, Rayp G (2005) Do consumers care about ethics? Willingness to pay for fair-trade coffee. Journal of Consumer Affairs 39:363–385

Stern PC (2000) New environmental theories: Toward a coherent theory of environmentally significant behavior. Journal of Social Issues 56:407–424

Stiglitz J (2002) Globalization and its discontents. Norton, New York and London

ZfB
ZEITSCHRIFT FÜR BETRIEBSWIRTSCHAFT

How Negative Environmental Impacts Drive MNE Exit: Stakeholder Dynamics in the DHL Europe Case

Elvira Haezendonck, Michaël Dooms, Alain Verbeke

Abstract: In 2003, DHL announced the planned hub extension at Brussels National Airport (BNA). During the ten months following, the societal debate on this expansion project metamorphosed into a trial of strength between the MNE (DHL) and various stakeholders, finally leading to DHL's hub exit from BNA. We interpret exit strategies using insights from stakeholder management and dynamics. We empirically analyze the changing levels of Mitchell, Agle and Wood's (1997) attributes, and the effects thereof on MNE exit. We argue that a stakeholder process approach may in some cases better explain exit decisions than the traditional financial or transaction cost arguments.

Keywords: Stakeholder theory · stakeholder dynamics · environmental impacts · MNE exit

JEL Classification: M14

Assistant Prof. Elvira Haezendonck, Ph.D. (✉)
Management and Strategy, Vrije Universiteit Brussel (VUB) and Solvay Business School, Pleinlaan 2, 1050 Brussels, Belgium
E-mail: elvira.haezendonck@vub.ac.be; http://www.vub.ac.be/BEDR/

Michaël Dooms, drs. (✉)
Research fellow Management and Strategy, Vrije Universiteit Brussel (VUB) and Solvay Business School, Pleinlaan 2, 1050 Brussels, Belgium
E-mail: michael.dooms@vub.ac.be; http://www.vub.ac.be/BEDR/

Prof. Alain Verbeke, Ph.D. (✉)
McCaig Chair in Management, University of Calgary and Vrije Universiteit Brussel (VUB) and Solvay Business School, Pleinlaan 2, 1050 Brussels, Belgium
E-mail: averbeke@vub.ac.be; http://www.vub.ac.be/BEDR/

1 Introduction

"If a primary stakeholder withdraws its support of an organization, the organization will cease to exist" (Clarkson, 1995: 106). The present paper relates this suggested dynamic process related to stakeholder involvement and its impact on MNE exit decisions, with both theoretical arguments and empirical, cased-based findings.

The paper explores the dynamic process of stakeholder salience and its effect on MNE strategy. The paper has three major sections. The first section describes the DHL case and its relevance for the paper. The second section discusses key theoretical constructs from the stakeholder management literature as regards stakeholder influences and the related potential MNE (des)investment strategies. Stakeholder dynamics are analyzed within the context of the Mitchell, Agle & Wood (1997) typology using the constructs of power, legitimacy and urgency which lead to salience and suggesting that the levels of these attributes may vary over time. In the context of this paper, MNE exit strategies are viewed as a sequential set of interrelated strategic decisions made by the firm, see Stafford and Watts (1990) and Stafford (1991). In section three, we analyze the DHL strategy during the stakeholder negotiation process and argue how this dynamic process explains the disinvestment decision of DHL. This potentially helps us to understand better why MNEs may favor exit, irrespective of the traditional sunk costs logic. The paper concludes with an agenda for further research.

2 Case description: the 'DHL super-hub location choice'

In the course of 2003, the multinational logistics provider DHL announced its plans to concentrate all its daily air cargo and express flights, including its large set of night flights in one European location. DHL provided two main reasons for this concentration strategy. First, in 2003 DHL was the only global express courier company (the other global firms being UPS, TNT and Federal Express) working with a system of several sub-regional hubs, dispersed throughout Europe. DHL assumed that the regional concentration of activities in a single European super-hub was likely to bring substantial logistical cost savings. Second, the handling capacity at Brussels National Airport (BNA), which served as the main hub in the regional system, was expected to reach its saturation level in 2007, thereby imposing substantial new investments in handling capacity. Therefore, DHL sent a formal request for capacity expansion proposals to several European airports, including Brussels National Airport.

From interviews with company representatives, as well as information provided through the media, three airports were – after a European Request for Proposal was sent to several airports and a business case analysis was carried out for a hub in these airports – shortlisted by DHL for its expansion plans. First, there was Brussels National Airport, which served as the main hub at the outset of this decision process. DHL has been historically present in this airport, its main advantages being the excellent weather conditions (BNA has one of the lowest number of closing days per year due to bad weather), the proximity to the market and the highly skilled and productive workforce. The main disadvantage of BNA was the growing stakeholder opposition from local communities through action committees and NGOs, as well as local and regional governments. BNA is located in a densely populated area, which led to a strong case for the restriction of night flights, and as a result, there was an agreement among various stakeholders that a ceiling had to be put on the development of DHL at BNA.

Second, Vatry airport, in the Champagne region in the North-East of France, located at 300 kilometres south of BNA, was shortlisted. The main advantages of this airport were the unlimited use of capacity at night and its proximity to the market from a European perspective, although it is physically located in one of the less densely populated areas in Europe. The main disadvantages are the closure of the airport for an average of 20 days a year due to fog as well as the social climate, including potential difficulties to find qualified labour.

The third alternative, for which a business case was developed by DHL, was the airport of Leipzig in former East-Germany. The main advantages were a relative limited restrictions on night flights, the presence of a workforce (though less skilled and productive as the Brussels' workforce), given relatively high unemployment rates in the region, and a favourable institutional context including the possibility to receive grants and subsidies from several governments. The main disadvantage of Leipzig was the remote location, as it is not located in the centre of economic activity in Europe at the moment. As a result, the distances to be covered by air transport would increase.

After further negotiations, the Vatry alternative was eliminated and the choice became one between BNA and Leipzig, with an initial stated preference from DHL to negotiate with BNA for an extension of the existing hub.

If the Pan-European super-hub were to be located at BNA, this would imply a dramatic increase in night flights (from about 13,000 DHL night flights a year in 2003 to approximately 25,000 – 35,000 in the future, depending on the growth scenario). The DHL request for a proposal for new capacity at BNA was positively highlighted in the national press in a period of decreasing economic growth and a downsizing trend in firms, especially given the large number of additional jobs that would be created in the wider region in case of a DHL expansion in Brussels (Sleuwaegen et al, 2003). The employment data reported in the press were provided by the firm itself and by the airport infrastructure manager (BIAC – Brussels International Airport Company). Immediately after the public announcement of the planned extension, various public agencies started reflecting on the expected societal costs (increasing number of night flights) and benefits (employment and value added) of such a development. Due to a prior, fragile political agreement reached on the night flights schedule (in terms of allowed number of flights and their routes), just months before DHL announced its extension plan, the planned expansion received substantial media attention. At the same time, pockets of resistance against the DHL expansion arose in the local communities surrounding the airport, based on the alleged negative effects of night flights on human health, and the resulting, increased healthcare costs, as well as lower worker productivity caused by a lack of peaceful sleep.

In January 2004, the Belgian Federal Government announced that it would respond to DHL's request for capacity expansion in Brussels by September 2004. This response was supposed to include, in principle, a feasible extension plan at Brussels National Airport. In case a capacity expansion at Brussels National Airport would not be possible, an alternative location on Belgian territory would be proposed. This latter option quickly disappeared however, as various feasibility analyses demonstrated that the costs to build a new airport or to transform existing regional or military airports were too high. In addition, the administrative hurdles to overcome in order to allow DHL to function effectively in one of these alternative locations on Belgian territory were considered excessive in each case.

A capacity extension at BNA proved to be the only feasible location in Belgium. However, such an extension also meant that the environmental permit given to the airport operator BIAC, set at maximum 25,000 night movements per year, would need to be renegotiated with the Flemish regional government, responsible for delivering environmental permits, in order to allow for more night flights. Furthermore, the Brussels regional government also needed to be involved, as each regional government determines the maximum noise levels tolerated on, or above, its territory. Though BNA is physically located on the territory of the Flemish Region, the airplane flight paths penetrate the Brussels Region, and a substantial number of night flights cross Brussels territory. As a result, the noise levels set by the Brussels Government would determine to a large extent the feasibility of the extension at BNA. The above suggests the need for an integrated policy approach by the Federal Government, and the two other regional governments involved. In practical terms, the Federal Government acted as the sole political negotiator with DHL, but taking into account that each decision it would take would also need to be approved and implemented by the Regional Governments involved. Therefore, substantial decision making power was actually held by the Regional Governments, namely the Brussels Government in terms of setting the maximum noise levels and determining the maximum number of night movements above the Brussels region, and the Flemish Government, responsible for delivering the airport environmental permit.

The above decision-making complexities meant that in the time span of a few months, a number of important, interrelated decisions with long-term impacts had to be taken by different governments. Here, a number of partially conflicting objectives could be observed among the various stakeholders, namely:

- DHL's objective to function within the context of a stable regulatory framework that would allow the company to organize a sufficient number of night flights, as required by its business plan, thereby ensuring long-term growth, especially given the important capital investments in infrastructure to be undertaken by the company.
- The airport operating company's (BIAC) objective to achieve the same, stable regulatory framework, and to keep the extended DHL-hub as a component of the airport's location advantages in attracting new cargo and passenger operators.
- The local community's objective to limit or eliminate entirely the DHL night flights, given the alleged health costs associated with night flights. The local community was represented by action committees, as well as elected representatives in the local municipal councils.
- The objectives of the different Governments involved that included economic growth, reduction of ecological harm and a variety of policy objectives.

In order to support the decision process, two studies were commissioned by the Province of Flemish Brabant, responsible for giving advice to the Flemish Government on the environmental permit for the airport activity.

- A study on the economic benefits of the extension project;
- A study on the health effects (including health costs) of the extension project.

Both studies led to the identification of a number of economic benefit and cost categories, including both quantifiable and non-quantifiable impacts. The parameters considered relevant and important were different for each stakeholder group, thereby imposing a stakeholder based evaluation process. The project was highly controversial and there was a clear perception that decisions would not be accepted if stakeholders were not involved. In practical terms, the study required stakeholder input from DHL itself, from the airport operator BIAC, and from the 'Government'. The complexity of the Belgian institutional model, with several governmental levels involved, turned out to be a key issue in this process.

In sum, during the ten-month period following the DHL announcement of its planned extension at BNA (December 2003), the debate metamorphosed into a trial of strength between a large MNE (DHL) and a wide variety of stakeholders, including government agencies at various levels, and several pressure groups.

On October 20th, 2004, DHL announced publicly that its BNA extension plan had to be canceled, due to a lack of government support. This cancellation also implied DHL's exit from BNA, at least as far as hub activities were concerned. This exit was expected to lead to a direct loss of 1,700 jobs by 2008, as opposed to a gain of 1,400 direct jobs with the extension. A few weeks later, DHL announced that the planned extension would – instead – take place at the airport of Leipzig, Germany, with 3,500 new jobs being foreseen. Leipzig would also receive the status of European hub.

The issue of expanding airport related activities must be assessed in the context of rapid, worldwide air traffic increase and the challenge to find suitable locations for the sustainable growth of the related activities in airport hubs. Public policy makers, as the guardians of the principles of sustainable development, must carefully weigh the various socio-economic and larger societal impacts of such location decisions. They must also attempt to achieve at least some consensus among all stakeholders involved, so as to avoid subsequent problems in the implementation stage of the MNE's (adapted) project. However, in the DHL case, the dynamics of the interaction process between the MNE and the various stakeholders involved, ultimately led to a complete exit of hub activities in Brussels. Importantly, in this case a number of government agencies switched from an initial position of favoring expansion to the preference of non-implementation, while DHL consistently favored the extension at BNA, until the political unfeasibility thereof had become clear. Considering the various stakeholders involved and the change in entry and exit preferences observed during the stakeholder interaction process, it is interesting to discuss this issue from a theoretical perspective (section 3).

3 Theoretical perspectives on stakeholder dynamics and MNE exit strategies

3.1 Stakeholder theory

In the DHL case, a stakeholder perspective is particularly interesting as the case involved several stakeholders with a variety of conflicting objectives.

Drawing on the seminal book "Strategic Management: A Stakeholder Approach" of Edward Freeman (1984) and on Donaldson and Preston (1995), stakeholders can be narrowly defined as any group or person that has a stake in the organization. Other actors, such as the media, might influence the organization, but do not have a direct stake in it (Donaldson and Preston, 1995). A narrow perspective on stakeholders – as opposed to a broad one (for example Freeman, 1984 and Starik, 1994) – enables organizations, to concentrate on just a few critical stakeholders. However, a stakeholder management approach is ultimately only useful to decision makers if it provides better insight into who can influence decision making processes and organizational outcomes in a substantive fashion, meaning that it is important to adopt a broad approach as to whom should be included as relevant stakeholders.

In this paper, we build primarily upon Mitchell, Agle and Wood (1997), who analyzed the salience of external stakeholders using the constructs of power, urgency and legitimacy. In their paper, these authors describe how stakeholders can be prioritized in terms of their ability to influence the actions of an organization (*power*), the degree to which stakeholder claims call for immediate attention (*urgency*) and in terms of their social acceptance and their expected structures and behavior (*legitimacy*) (Mitchell et.al., 1997). These three concepts enable the analysis of stakeholder salience or determining which stakeholders should get most attention from decision makers in the organization. Any high score on only a single one of these attributes does not guarantee high salience of a stakeholder. Mitchell et.al. (1997) importantly notice that these attributes are subjective and may, whether or not consciously, not lead to expected behavior. This subjectivity and behavior may also change in time. This already suggests that stakeholders can gain or lose salience to decision makers in an organization.

3.2 Firm exit rationales and process

Firm boundaries are rarely, if ever, static for an extended period of time. Rather, growth is sought through internal or external means and thus, activities are expanded and new markets are entered. At the same time, however, (and sometimes by the same firm in the same timeframe) activities may be divested, and markets or product lines may be let go.

Essentially, firm exit has four rationales which may but need not occur simultaneously: financial, strategic, organizational and forced exit. According to Boddewyn (1979), financial considerations are most important when dealing with firm exit, which may occur to divest unprofitable business or to obtain funds for new investment (see also Karakaya, 2000; Dewitt, 1998). This rationale thus groups all firm-level exit types aimed at reducing financial losses, or at obtaining funds that could be allocated more profitably elsewhere. The second rationale groups all firm-level exit types associated with strategic considerations, i.e. in case of a lack of internal fit among the various activities a firm carries out (Haynes, Thompson and Wright, 2003). A third reason is organizational, whereby firm-level exit is due to a firm becoming too large to manage (Boddewyn, 1979). Finally, firms may exit a certain activity if they are forced to, for example due to government action which can range from expropriation or nationalization to competition policy (Kobrin, 1984; Minor, 1994).

In the DHL case, we can assume that financial, strategic, nor organizational reasons primarily led to the exit decision in BNA. BNA, as part of the three shortlisted locations for DHL's hub, represented a strong and valid business case for DHL; first, an initial Request for Proposal was sent to a large sample of European Airports; then, three airports, which matched DHL's criteria for a valid business case, were shortlisted and intense negotiations with the shortlisted airports to develop a refined business case and implementation plan for each location were started. In-depth analysis of company information, including interviews with representatives and press announcements, did not suggest that a financial, strategic or organizational aspect could be considered as the ultimate or single rationale for exit in this case. The rationale for DHL's exit decision from BNA would rather be linked to the forced exit argument, where stakeholder dynamics, as argued in section 4, better explain DHL's exit decision.

It should be noted that in the situation described in this paper, DHL engaged simultaneously in partial exit from Belgium and entry into Germany, as far as the regional hub activities are concerned. In fact, DHL's exit from Belgium would not have occurred if no alternative were available. We should therefore consider the two decisions (exit from Belgium and expansion in Germany) as inter-dependent. Again, literature on this kind of strategic action is limited. For example, crucial in the literature that considers the MNE as a network is the notion of uncertainty (Kogut and Kulatilaka, 1994). Although strategic flexibility allows a firm to shift resources and thus simultaneously enter or exit markets, forced divestment as in our case cannot be considered as the source of a conventional source of uncertainty that could have yielded a benefit. More specifically, rather than benefiting from an option, the firm in our case does not only lack a strategic option to shift resources, but it is actually forced to consider divestment. To our knowledge, Stafford and Watts (1990) represents the only scholarly analysis focused on how a firm's local environment can have some impact when considering selective closure, whereby an MNE chooses among similar plants those that are to be divested. However, even those authors find that environmental factors such as poor labor relations are not dominant when considering the top 10 differences between closed and surviving plants, while the parameter 'unfavorable local government attitudes' did not even feature as a factor in this top 10. In this context, the authors state that "it is important to reiterate that the approach … did not ask managers to identify factors they considered in deciding which plant to close" (Stafford and Watts, 1990: 437).

Despite various reasons fostering exit, firms seldom exit a market as soon as problems occur, except in the case of nationalizations. The primary reason for this observation is that firm exit is not a singular action but rather a process, as noted above (Boddewyn, 1983; Duhaime and Schwenk, 1985; Clark and Wrigley, 1997; and Pauwels and Matthyssens, 2003). Exit is generally viewed as a sequential set of interrelated strategic decisions made by a firm. As a result, firm exit is the last remaining step after various other options have been tried and have met no success (Clark and Wrigley, 1997). In addition, the process itself is subject to various issues such as sunk costs (Clark and Wrigley, 1997), the degree of institutionalization of the activity, and the degree of endogenous and exogenous stress experienced by managers in the firm (Pauwels and Matthyssens, 2003). A second reason why firm exit may be delayed – or not actually implemented – is strategic in nature: either the activity is intimately related to the firm's core activities through backward or forward linkages (Karakaya, 2000), or the firm wants to continue to enjoy the benefits of mutual forbearance by being active in the same markets as its competitors (Boeker, Goodstein, Stephan and Murmann, 1997).

3.3 Exit process from a stakeholder perspective

Mitchell et. al. (1997) did explicitly suggest that the attributes power, legitimacy and urgency may be variable or transitory. If the concepts power, urgency and legitimacy change during the course of a specific period, this may affect stakeholder salience. If a decision-maker, whether in the firm or in government, fails to recognize the changing dynamics when applying the typology of Mitchell et. al. (1997), this may result in, for example, the underestimation of the power or urgency of a critical stakeholder. In this case, the stakeholder may believe that his expectations are no longer met and, as a result, he may well withdraw his support for a firm's project. This could lead to a divestment or exit decision by the firm (Clarkson, 1995). Indeed, Clarkson (1995, p. 112) argues that when a firm fails to achieve the balance in meeting stakeholder expectations and fulfilling its responsibilities to its primary stakeholder groups, this may result in the withdrawal of support of a stakeholder group and ultimately threaten the firm's survival. Here, Clarkson suggests that when an imbalance between stakeholder and shareholder preferences is persistent, this can lead to corporate closure. Whereas the traditional exit strategy literature points to the firm as the most common decision maker as regards a divestment or exit strategy (based on for example sunk costs, see Rosenbaum and Lamort, 1992; Clark and Wrigley, 1997), stakeholder theory (Freeman, 1984; Clarkson, 1995; Wartick, 1994; Wood, 1994; Mitchell, Agle and Wood 1997) suggests that the pressure or lack of support from a critical stakeholder, may – at least in some cases – better explain why firms engage in closing plants or engage in divesting activities. Exit decisions are traditionally linked with financial, corporate governance or transaction cost arguments (such as described in Haynes, Thompson and Wright, 2003). But Haynes, Thompson and Wright (2003) also suggest that further empirical research may be required to explore the impact of the environment, rather than focusing on governance mechanisms to explain corporate restructuring, though both are undoubtedly interrelated. We argue in this paper that stakeholder theory may complement transaction cost theory in explaining firm's exit decisions. Although a few scholarly articles have recognized the need to interpret exit strategies using insights from stakeholder management models (for example Haynes, Thompson and Wright, 2003), no empirical research actually recognizes the dynamics of stakeholder management, and the effects thereof on MNE exit decisions.

4 Stakeholder dynamics explaining exit: a case-based analysis

In the DHL case, described in section 2, the dynamics of the interaction process between the MNE and the various stakeholders involved, ultimately led to a full-fledged exit decision. Why and how did this happen?

Importantly, in this case a number of government agencies switched from an initial 'entry' preference (i.e. favoring expansion) to an 'exit' preference (i.e. favoring non-implementation, and the related loss of hub status for BNA), while DHL consistently favored the 'entry' approach until the political unfeasibility thereof had become clear.

In order to explain the shift in preference and the final exit decision, we developed a framework consisting of two dimensions, namely (1) the strategy of the MNE, which could be one favouring entry (expansion) or exit and (2) the strategy of government agencies, which could also build upon an entry (expansion) or exit preference. As a result, a two by two matrix with four quadrants is created (see Figure 1).

Quadrant 1 represents a rather straightforward case whereby both actors share a preference for entry (expansion). This is particularly the case e.g. in emerging countries, where governments actively try to attract large industrial activities to support their economic development (e.g. through favourable tax regimes, infrastructure development) and multinational companies try to enter stimulated by the presence of favourable markets or factor conditions (such as cheap labour, the presence of natural resources, etc.).

Quadrant 2 shows a situation whereby the MNE pursues entry (or expansion if the MNE is already active in the country), but with government strategy favouring exit or non expansion (e.g., in case it does not want a foreign MNE to increase investments). This could be the case, e.g., when government refuses to grant a license to operate (e.g. environmental or other permits) to an MNE with investment plans in the country. Various reasons could underlie such a situation, such as (1) the activity of the MNE causes negative environmental externalities and the investment project or existing activity of the MNE does not show sufficient effort to reduce the externalities (2) the activity of the MNE harms the development of host country firms from a market perspective, and government wants to protect these firms from bankruptcy. Other examples include a situation whereby the MNE is active in a sector which suddenly becomes of national strategic importance for the host country (e.g. services or technology linked to national security issues, energy or transport) and whereby government wants to be in control of the sector or at least keep it in national hands. In these cases, exit or non entry from the viewpoint of the MNE is non voluntary and leads in most cases to a loss of efficiency at the micro-level.

From the viewpoint of governments in advanced economies, this situation in quadrant 2 has become increasingly unrealistic, as economic policies are strongly oriented towards the creation of employment. This normally results in a favourable attitude of governments towards foreign investors, with most countries competing with each other to attract these footloose investment projects. This attitude is confirmed by quadrant 3, where the MNE pursues an exit strategy process, and government wants to prevent this exit and adopts an entry strategy (or non exit). This is a very familiar case in Western economies, in particular in industrial sectors such as textiles and automotive. Announcements of MNEs downsizing plans in Western economies (which can be part of a relocation strategy towards emerging economies) are in most cases immediately followed by government action to maintain the activity in the short term by e.g. by making more flexible labour and fiscal regulations (e.g. Belgium changed labour and fiscal laws concerning team labour to prevent the exit of automotive firms), giving direct subsidies and grants and/or financing infrastructure that benefits the firm, reviewing environmental permits in favour of the MNE, etc..

Finally, quadrant 4 shows a rather specific situation with both parties simultaneously pursuing an exit strategy process. This could happen if the MNE wants to exit from a spe-

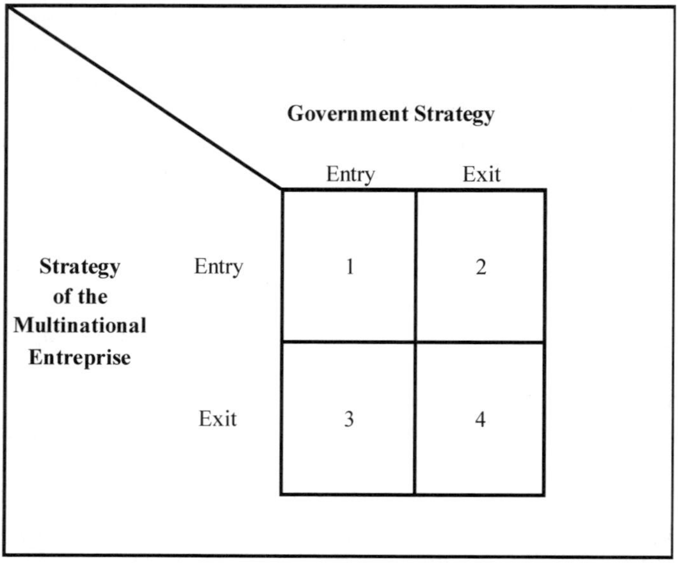

Fig. 1. Potential strategies of government and MNE during a foreign direct (dis)investment process
Source: Dooms, M., A. Verbeke, C. Macharis and R. S'Jegers (2006) ***De zaak DHL***. Garant Publishers, Leuven, Belgium.

cific region or country (e.g. for market and/or financial reasons) and the concerned governments see the opportunity to develop local firms and/or attract other foreign investors.

To gain insight into the process of exit in the DHL case, we apply the framework presented in Figure 1. To perform this analysis, we analyzed all publicly available press releases, spanning the period between the moment in time when DHL and the government first expressed they favoured the entry option (December 2003) and the moment when both parties almost simultaneously announced the exit of DHL at separate press conferences. We also gathered inside information during our involvement in a socio-economic evaluation study for the Province of Flemish Brabant, where we were able to work with representatives from government agencies, the firm DHL and BNA representatives.

Our analysis led to the conclusion that between December 2003 and September 2004 the situation of quadrant 1 applied, as the Belgian Federal Government had examined the request of DHL and engaged itself to create the conditions to realize the project. From its side, DHL examined the possibility to grow at BNA reviewing its forecasts and participating in a large number of joint workshops with BNA and government representatives. A special negotiator was assigned by the Federal Government to lead the workshops and different taskforces. A final decision was to be reached by the 1st of September 2004. During the period preceding this deadline, a technical agreement between the negotiators was reached, but this was not made public, as it appeared difficult to reach a political agreement (in fact, the existence of this technical agreement was only confirmed end of October 2004). The difficulties to reach a political agreement were apparently caused by the election results of June 2004, whereby powers shifted in the regional governments, leading to different political coalitions governing on the federal level and the regional level.

As the Regional Governments needed to approve the environmental permit and set the noise regulations, they became important decision makers. The asymmetry of political coalitions started to play its role as it was apparent that the tradition of loyalty of the regional level towards the federal level for dossiers of national economic significance was vanishing. Although there is no solid factual information that this election and its consequences played such an important role, it is noteworthy that the deadline of the 1st of September, mentioned above, elapsed though a technical agreement negotiated by Federal Government representatives existed. DHL declared that it wanted a decision in the short run (with no deadline), but that it remained confident that an agreement could be reached.

Three governments were directly involved in this dossier, of which two – the Federal Government and the Flemish Regional Government – gave their approval to the technical agreement for expansion on the 7th of September. The 11th of September, the Brussels Regional Government declared, however, that it would not approve the expansion, even if this meant that the creation of jobs for the large pool of unemployed workers in the Brussels region would not take place, as the regional coalition agreement did not allow for an increase of night flights. This meant that all additional negative externalities would *de facto* be concentrated above (and on) Flemish regional territory. Almost immediately after the decision of the Brussels Regional Government, a number of Flemish Government Ministers expressed their view that an exit scenario was now a possibility. At this point, Government (in this case seen as the sum of different governmental levels with decision power) altered its position from a preference for entry (expansion) towards exit. DHL confirmed its continued preference for expansion in Belgium but also stated that a decision was urgent. As a result, the situation in this stage became consistent with quadrant 2 in Figure 1.

During the period between 20 and 27 September 2004, the three governmental levels negotiated but were not able to reach an agreement, which implied that the entry preference strategy of Government (as the sum of the three Governments) did not hold anymore, as the Brussels Regional Government explicitly decided in favour of exit, further confirming a quadrant 2 situation had been reached. A final blow to the project came when the Governments claimed that a technical detail, i.e. the number of flights to be allowed with a specific type of noisy plane (MD-11), was not clear in the agreement, and disagreement on this issue was voiced between the number proposed by government and the number proposed by DHL. This resulted in a demand from all governments to DHL to further examine the use of the MD-11 (which at the time, had no substitute for intercontinental cargo flights). Two observations are important: (1) there were no public declarations from any party involved during the first two weeks of October, whereas the project had featured prominently in the national media almost every day in September and (2) suddenly, the three Governments acted as one actor with one voice. Several negotiators declared ex-post that they had the feeling that the issue of noisy MD-11's had been misused to allow the Governments and the firm to gain time and prepare a joint exit strategy which would lead to (1) salvaging the Federal Government's reputation/credibility, as the prime minister had linked his mandate to success in the DHL expansion case, and (2) limiting the negative fall-out on the image of the firm DHL. These observations suggest that both parties simultaneously started to pursue exit from that point on.

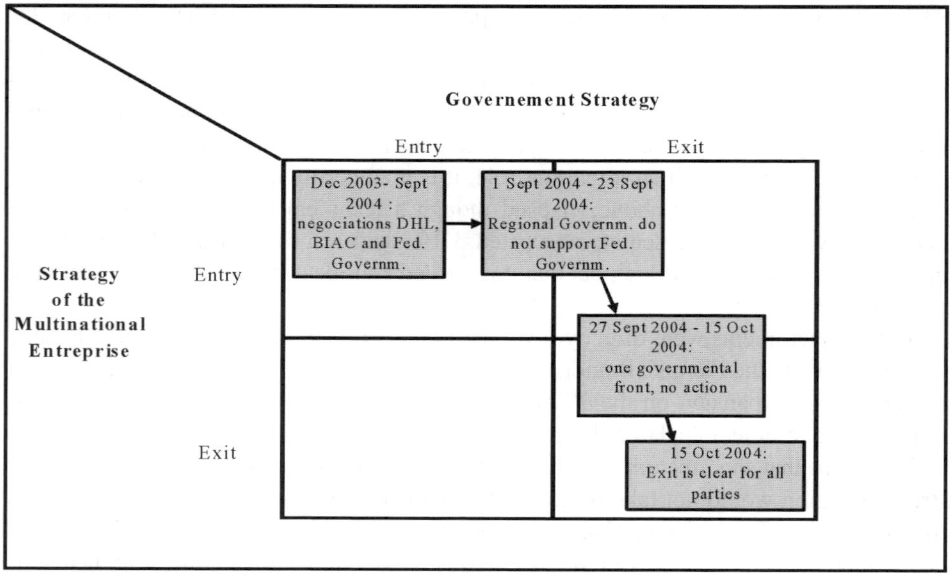

Fig. 2. Dynamic presentation of the process in the DHL case
Source: Dooms, M., A. Verbeke, C. Macharis and R. S'Jegers (2006) ***De zaak DHL***. Garant Publishers, Leuven, Belgium.

On October 15th, the (Federal Government's) Prime Minister publicly said that an expansion scenario would be "difficult" to implement, meaning that the Federal Government, the initial sponsor of the project, had also shifted towards a preference for exit. Media silence at DHL also suggested that the MNE had shifted towards an exit strategy and started preparations to implement this decision. On October 20th, a press conference with the three leaders of the different Governments was held, explaining that all government levels considered the project as non-feasible and as a result, would not create the conditions for an expansion. Simultaneously, DHL held a press conference to announce that an expansion in Brussels had no societal acceptance, and that the European Enterprise Council had in the meanwhile decided not to expand in Brussels. On November 9th, the firm announced the move of the intercontinental hub to Leipzig from 2008 on.

First, the analysis of the process shows that the position of both the MNE and government in terms of entry or exit strategy is in some cases difficult to determine and subject to change during a negotiation on the realisation of a foreign direct investment project. The case study suggests that both parties moved their preferences from quadrant 1 to quadrant 4 via a short period in quadrant 2, see Figure 2. As suggested earlier, quadrant 4 is a specific situation, which occurs rather rarely in international business, but which we expect to gain importance in the future, as e.g. external stakeholders such as NGOs and other interest groups put pressure on governments and MNEs to adopt environmental standards and regulation, which are incompatible with the efficiency of commercial or industrial activities.

Second, the application of the framework to the DHL case at Brussels National Airport confirms the insights from the extant literature that firm exit is not a singular action but rather a process (Boddewyn, 1983; Duhaime and Schwenk, 1985; Clark and Wrigley, 1997; and Pauwels and Matthyssens, 2003) and that exit is generally viewed as a sequential set of interrelated strategic decisions made by a firm. As a result, firm exit is the last remaining step after various other options have been tried and have met no success (Clark and Wrigley, 1997). The behavior of DHL clearly points in this direction as the firm started with an open request for proposals addressed to several airports, brought the long list down to a shortlist, and continued with one preferred partner (BNA) and with one alternative option (Leipzig). Evidence from the process in this case shows that the exit decision came rather late and that substantial resources where invested to examine the different options at BNA. The fact that the expiration of the 1st September deadline did not lead to an immediate exit decision (despite a technical agreement) and that negotiations were held until the beginning of October, shows that the firm considered exit the last remaining option.

Third, the case also strongly supports the argument that stakeholder theory (Freeman, 1984; Clarkson, 1995; Wartick, 1994; Wood, 1994; Mitchell, Agle and Wood, 1997), or the pressure or lack of support of a critical stakeholder, may better – or at least as a complement to other exit drivers such as sunk costs – explain why firms engage in closing plants or even impose divestment. Here, the exit decision was to a large extent based on the lack of support from Government due to the environmental and health effects of night flights. During the negotiation process, the existence of health effects was continuously and strongly communicated to the different governments and media by several NGOs as well as local community interest groups and this communication strategy clearly influenced and put pressure on all political parties as well as key politicians, as the constituencies of some Ministers were located within noisy areas around the airport.

Complementary evidence that stakeholder theory plays an important role to understand the dynamics of the exit process, can be found in the application of the typology of Mitchell et. al. (1997) to the DHL exit case.

In order to apply this framework, we first need to acknowledge that there are in fact two focal organizations in this case, namely the MNE DHL and government. We then need to make a distinction between two types of dynamic aspects that can play a role. First, during the process, a stakeholder can lose or gain with respect to certain attributes of the Mitchell et. al. (1997) framework, as perceived by the focal organization. We call this a type 1 dynamic aspect. Second, there can be a gap between how a specific stakeholder is expected to behave by the focal organization, or in other words, which attributes of this stakeholder are expected to show up in a particular decision making process on the one hand, and the behavior or attributes the stakeholder exhibits in reality on the other hand. This gap then calls for an adaptation of stakeholder management by the focal organization. We call this a type 2 dynamic aspect. The difference between the two dynamic aspects is that in the first case, something changes during the process (the importance of certain attributes of a stakeholder changes), whereas in the second case, expected attributes of a stakeholder are not present and/or unexpected attributes are present. The implication for decision-makers involved in complex projects is to try to gain insight ex-ante into the likelihood and consequences of the occurrence of such dynamic aspects.

With regard to the type 1 dynamic aspects, two examples can be identified in the DHL case. With regard to DHL as a stakeholder, with government being the focal organization,

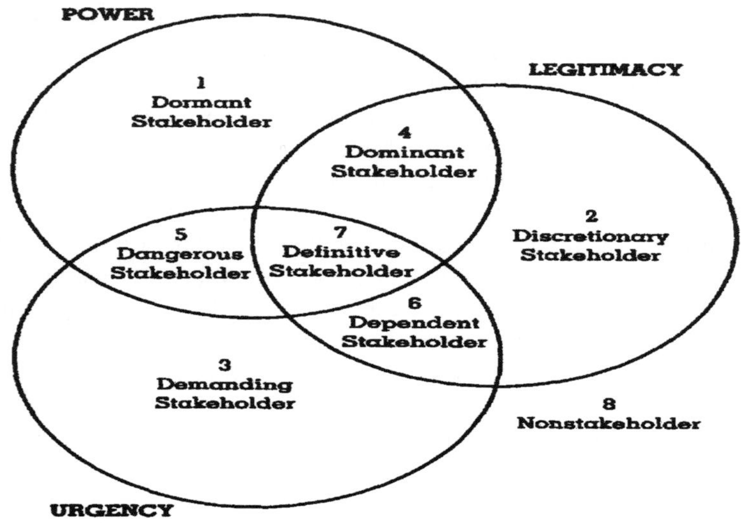

Fig. 3. Stakeholder typology: one, two, or three attributes present
Source: Mitchell, R.K., Agle, B.R. & Wood, D.J. 1997. Toward a theory of stakeholder identification and salience: Defining the principle of who and what really counts. ***Academy of Management Review***, 22(4): p. 874.

it can be argued that at the beginning of the project, the attributes power and legitimacy were present, whereas urgency was less present, defining DHL as a dominant stakeholder from the government's perspective. It is clear that DHL could influence the behavior of Government and BIAC as both committed substantial resources to examine the possibility of an expansion on the Belgian territory. In terms of urgency, though both Government and BIAC took immediate action, DHL did not explicitly set or impose deadlines on other stakeholders and allowed other parties to take their time in the process (for example the installation of task forces to reach a technical agreement and the deadline of September 1st, 9 months after the announcement). This attribute changed on September 1st, when no decision was taken by Government and DHL called for a short term solution. Also, DHL's power decreased as an increasing number of government members publicly declared they did not believe in the project and that government should develop plans to grow and support other activities. As a result, DHL moved towards being a dependent stakeholder from the perspective of government as the focal organization.

The second example concerns Government as a stakeholder, with DHL as the focal organization. The government initially exhibited all attributes of a salient stakeholder. The government exhibited power, as it would ultimately decide on the environmental permit, which eventually would have made the project feasible. The urgency was present as the objective of the government was the creation of 200,000 jobs during its legislation, and the DHL project was considered as a symbol of that objective. As a result, Government called other stakeholders (e.g. BIAC) and the different government agencies to immediate action. Legitimacy was present, as it was initially agreed by all political parties and governments that the Federal Government would be the unique partner and decision maker in the negotiation. As a result, Government was a definitive stakeholder from the perspective of DHL as the focal organization. This changed after the deadline of September 1st and certainly after the failed inter-governmental negotiation of end of September 2004. First, legitimacy was lost after the unified governmental front was breached by the Brussels Regional Government, and again when during the course of events, a growing number of government representatives questioned the expansion and moved from a preference for expansion to exit. Second, the attribute of urgency was also lost after the failure of inter-governmental negotiations as the three governments considered the letter with the demand for information as the last action and did not put forward a deadline any more. As a result, government changed from being a definitive stakeholder to a dormant stakeholder from the perspective of DHL as the focal organization.

The dynamics of the process and the changing attributes show that DHL and the Government, the two main decision makers in the negotiation, changed from respectively a dominant and a definitive stakeholder towards a dependent and a dormant stakeholder, in each case from the perspective of the other party as the focal organization. These changing attributes contributed to dynamic relation between the parties: besides a reversal in terms of urgency, one stakeholder lost the attribute of power (DHL) and another stakeholder legitimacy (the Government, as it did not speak with one voice any more). Without these changes in attributes, both parties would probably have stayed in quadrant 1 and entry would have been a possible outcome.

With regard to type 2 dynamic aspects, two examples can be identified. First, local community interest groups (to a lesser extent NGOs) are normally expected to be demanding stakeholders, irrespective of whether government or DHL is the focal organization. The urgency attribute was present since a few years, because these interest groups help and encourage their members to continuously file lawsuits, imposing government to devote immediate attention. The legitimacy of local community interest groups was expected to be low as in this case, they did not represent a large number of the total population and sometimes even file lawsuits against each other's interests (especially on the issue of geographic dispersion of flights). However, it turned out that these interest groups had strong legitimacy in this case as they (1) spoke in univocal terms against the expansion plans of DHL, (2) they used modern communication means and the media very effectively (e.g. e-mails to politicians, lots of media presence) and (3) they were able to mobilize experts (doctors, engineers, lawyers, etc.) for their case as spokespersons. The power attribute is normally not present, because the interest groups involved had not been able to influence directly the actions of Government in the past (e.g. regarding flight dispersion plans). However, in this case, their pressure on government members led to a strong focus of the governmental discussion on environmental and noise impacts, and less on the positive impact of employment.

We can add a second example of type 2 dynamics. Trade unions, again whether viewed from the perspective of the firm or government as the focal organization, usually play a definitive role as stakeholder in such cases; they define the social climate, are a partner in negotiations for changes in social law and labor conditions and can call for immediate action through strikes. Furthermore, their representatives are considered legitimate because they are elected through social elections, and they are invited to collective bargaining. In this case, with several thousand of jobs at stake, the trade unions showed discretionary stakeholder behavior because both urgency and power were not present. Only at

Tab. 1. Summary of dynamic stakeholder aspects as applied to the DHL case

Stakeholder	Power	Urgency	Legitimacy
DHL (government as focal org.)	From High to Low	Low to High	High
Brussels International Airport Company	High	Low	High
Government (DHL as focal org.)	High	From High to Low	High to Low
Local Communities and NGOs	*Low* (High)	*High* (High)	*Low* (High)
Trade Unions	*High* (Low)	*High* (Low)	*High* (High)

In italics: expected attributes.
Between brackets: actual/realized attributes.

the beginning of October, when a concrete threat of exit became apparent, relatively small protests were held in the centre of Brussels. Only after the exit decision was announced, the trade unions were invited by Government, but mainly to guide the social consequences of the exit process. Whereas the local community interest groups had been continuously active since the negotiations started, the trade unions only surged when the deadline of September 1st passed and signs of a potential exit of DHL became apparent. Under normal circumstances, the trade unions would have been expected to balance the discussion between ecological and economic objectives. In fact, their influence on Government was expected to be larger as they were legitimate actors in the process, and had strong means to create urgency (strikes usually create more "urgency" than lawsuits).

In this case, the gap between expected behavior and actual, realized behavior of external stakeholders involved in the decision process, such as local communities and trade unions also pushed the Government more towards exit. This process towards exit was reinforced as a traditionally less powerful and less legitimate stakeholder actually exhibited these attributes (the local community) and vice versa (the trade unions which lost power and urgency). Table 1 provides a summary of the above analysis.

The overall conclusion of this case study is that managers involved in complex projects, in particular entry and exit processes of MNEs, should not only take into account static stakeholder theory, as articulated by Mitchell et al. (1997), but also the often complex and dynamic aspects associated with stakeholder behavior, both related to the changing nature or importance of salience attributes and to the potential gap between expected and real behavior as regards power, urgency and legitimacy.

We also expect the type of situation explained in the case-study, i.e., MNE projects with substantial positive employment effects being rejected and leading to exit of the MNE activity, to become more common in the future due to external stakeholders imposing or influencing Governments to setting increasingly strict operating conditions on large firms that will often be irreconcilable with efficient commercial and industrial activities. In such cases, this will simply mean MNE activity relocation in geographic space to more favorable jurisdictions where stakeholders are less concerned with perceived negative externalities.

5 Conclusions

In this paper, we have shown that the observed levels of power, legitimacy and urgency exhibited by various stakeholders, i.e., the key inputs for modern stakeholder management as articulated by Mitchell et al. (1997), are subject to change over time. In other words, an analysis of dynamic aspects of stakeholder salience attributes appears required. In the DHL case some of these attributes indeed appeared unstable, sometimes even within short time periods such as a few weeks or months. In other words, modern stakeholder management analysis should move beyond trying to identify static scores allegedly exhibited by various stakeholders for salience attributes, and take into account dynamic elements.

More importantly, our analysis suggests that in complex business – government negotiations, such as the dialogue on large-scale investment projects by foreign MNEs that are associated with substantial economic and environmental impacts, a dual stakeholder management analysis should be conducted, i.e., from the perspective of the two main focal organizations (government and the MNE involved). The need for such dynamic, dual stakeholder management analysis will clearly become more critical, as an increasing number of stakeholders external to the two focal organizations exhibit salience attributes, thereby increasing their ability to influence decision-making so as to pursue their narrow interests. The critical point here is that the salience of these external stakeholders may be different from the perspective of each of the two focal organizations, and this asymmetry may in turn alter the outcome of the dialogue between the two focal organizations. In the case of DHL, a joint preference for MNE expansion in Belgium was ultimately transformed into a joint preference for exit.

References

Agle BR, Mitchell RK, Sonnenfeld JA (1999) Who matters to CEOs? An -investigation of stakeholder attributes and salience, corporate performance, and CEO values. Academy of Management Journal 42(5):507–525

Boddewyn JJ (1979) 'Foreign divestment: Magnitude and factors'. Journal of International Business Studies 10(1):21–26

Boddewyn JJ (1983) 'Foreign and domestic divestment and investment decisions: Like or unlike?'. Journal of International Business Studies 14(3):23–35

Boeker W, Goodstein J, Stephan J, Murmann JP (1997) 'Competition in a multimarket environment: The case of market exit'. Organization Science 8(2):126–142

Clark GL, Wrigley N (1997) 'Exit, the firm and sunk costs: reconceptualizing the corporate geography of disinvestment and plant closure'. Progress in Human Geography 21(3):338–358

Clarkson MBE (1995) A stakeholder framework for analyzing and evaluating corporate social performance. Academy of Management Review 20(1):92–117

Cochran P (1994) The Toronto conference: Reflections on stakeholder theory (Essay by Phil Cochran). Business & Society 33(1):95–98

DeWitt R-L (1998) 'Firm, industry, and strategic influences on choice of downsizing approach'. Strategic Management Journal 19(1):59–79

Donaldson T, Preston LE (1995) The stakeholder theory of the corporation: concepts, evidence, and implications. Academy of Management Review 20(1):65–91

Dooms M, Verbeke A, Macharis C, S'Jegers R (2006) De zaak DHL. Garant Publishers, Leuven, Belgium

Duhaime IM, Schwenk CR (1985) 'Conjectures on cognitive simplification in acquisition and divestment decision making'. Academy of Management Review 10(2):287–295

Freeman RE (1984) Strategic management: A stakeholder approach. Boston: Pitman
Freeman RE (1999) Divergent stakeholder theory. (A response to Jones & Wicks, 1995). Academy of Management Review 24(2):233–236
Haynes M, Thompson S, Wright M (2003) 'The determinants of corporate divestment: Evidence from a panel of UK firms'. Journal of Economic Behaviour & Organization 52(1):147–166
Hill CWL, Jones TM (1992) Stakeholder-agency theory. Journal of Management Studies 29:131–152
Jensen MC (1986) Agency Cost Of Free Cash Flow, Corporate Finance, and Takeovers. American Economic Review, Vol. 76, No. 2
Jones TM (1995) Instrumental stakeholder theory : A synthesis of ethics and economics. Academy of Management Review 20(2):404–437
Jones TM, Wicks AC (1999) Convergent stakeholder theory. Academy of Management Review 24(2):206–221
Karakaya F (2000) 'Market exit and barriers to exit: Theory and practice'. Psychology and Marketing 17(8): 651–668
Kobrin ST (1984) 'Expropriation as an attempt to control foreign firms in LDCs: Trends from 1960 to 1979'. International Studies Quarterly 28(3):329–348
Minor MS (1994) 'The demise of expropriation as an instrument of LDC policy, 1980–1992'. Journal of International Business Studies 25(1):177–188
Mitchell RK, Agle BR, Wood DJ (1997) Toward a theory of stakeholder identification and salience: Defining the principle of who and what really counts. Academy of Management Review 22(4):853–886
Pauwels P, Matthyssens P (2003) 'The Dynamics of International Market Withdrawal' in Jain SC (ed), Handbook of research in international marketing, Edward Elgar Publishing
Preston LE, Sapienza HJ (1990) Stakeholder management and corporate performance. The Journal of Behavioral Economics 19(4):361–375
Reed D (1999) Stakeholder management theory: A critical theory perspective. Business Ethics Quarterly 9(3): 453–483
Savage GT, Nix TW, Whitehead CJ, Blair JD (1991) Strategies for assessing and managing organizational stakeholders. Academy of Management Executive 5(2):51–75
Shankman NA (1999) Reframing the debate between agency and stakeholder theories of the firm. Journal of Business Ethics 19(4):319–334
Starik M (1994) The Toronto conference: Reflections on stakeholder theory (Essay by Mark Starik). Business & Society 33(1):89–95
Wartick S (1994) The Toronto conference: Reflections on stakeholder theory (Essay by Steve Wartick). Business & Society 33(1):110–117
Williamson O (1996) Economic organization: The case for candor. Academy of Management Review 21(1): 48–57
Wood DJ (1994) The Toronto conference: Reflections on stakeholder theory (Essay by Donna J. Wood). Business & Society 33(1):101–105

GRUNDSÄTZE UND ZIELE

Die **Zeitschrift für Betriebswirtschaft (ZfB)** ist eine der ältesten deutschen Fachzeitschriften der Betriebswirtschaftslehre. Sie wurde im Jahre 1924 von Fritz Schmidt begründet und von Wilhelm Kalveram, Erich Gutenberg und Horst Albach fortgeführt. Sie wird heute von 14 Universitätsprofessoren, die als **Department Editors** fungieren, herausgegeben. Dem **Editorial Board** gehören namhafte Persönlichkeiten aus Universität und Wirtschaftspraxis an. Die Fachvertreter stammen aus den USA, Japan und Europa.

Die ZfB verfolgt das Ziel, die **Forschung auf dem Gebiet der Betriebswirtschaftslehre** anzuregen sowie zur Verbreitung und Anwendung ihrer Ergebnisse beizutragen. Sie betont die Einheit des Faches; enger und einseitiger Spezialisierung in der Betriebswirtschaftslehre will sie entgegenwirken. Die Zeitschrift dient dem **Gedankenaustausch zwischen Wissenschaft und Unternehmenspraxis**. Sie will die betriebswirtschaftliche Forschung auf wichtige betriebswirtschaftliche Probleme in der Praxis aufmerksam machen und sie durch Anregungen aus der Unternehmenspraxis befruchten.

In der ZfB können auch englischsprachige Aufsätze veröffentlicht werden. Die Herausgeber begrüßen die Einreichung englischsprachiger Beiträge von deutschen und internationalen Wissenschaftlern. Durch die Zusammenfassungen in englischer Sprache sind die deutschsprachigen Aufsätze der ZfB auch internationalen Referatenorganen zugänglich. Im Journal of Economic Literature werden die Aufsätze der ZfB zum Beispiel laufend referiert.

Die Qualität der Aufsätze in der ZfB wird durch die Herausgeber und einen Kreis renommierter Gutachter gewährleistet. Das **Begutachtungsverfahren** ist doppelt verdeckt und wahrt damit die Anonymität von Autoren wie Gutachtern gemäß den international üblichen Standards. Jeder Beitrag wird von zwei Fachgutachtern beurteilt. Bei abweichenden Gutachten wird ein Drittgutachter bestellt. Die Department Editors entscheiden auf der Grundlage der Gutachten eigenverantwortlich über die Annahme und Ablehnung der von ihnen betreuten Manuskripte. Sie können Beiträge auch ohne Begutachtungsverfahren ablehnen, wenn diese formal oder inhaltlich von den Vorgaben der ZfB abweichen.

Die ZfB veröffentlicht im Einklang mit diesen Grundsätzen und Zielen:

- **Aufsätze** zu theoretischen und praktischen Fragen der Betriebswirtschaftslehre einschließlich von Arbeiten junger Wissenschaftler, denen sie ein Forum für die Diskussion und die Verbreitung ihrer Forschungsergebnisse eröffnet,
- **Ergebnisse der Diskussion** aktueller betriebswirtschaftlicher Themen zwischen Wissenschaftlern und Praktikern,
- **Berichte** über den Einsatz wissenschaftlicher Instrumente und Konzepte bei der Lösung von betriebswirtschaftlichen Problemen in der Praxis,
- **Schilderungen von Problemen** aus der Praxis zur Anregung der betriebswirtschaftlichen Forschung,
- **„State of the Art"-Artikel,** in denen Entwicklung und Stand der Betriebswirtschaftslehre eines Teilgebietes dargelegt werden.

Die ZfB informiert ihre Leser über **Neuerscheinungen** in der Betriebswirtschafslehre und der Management Literatur durch ausführliche Rezensionen und Kurzbesprechungen und berichtet in ihrem **Nachrichtenteil** regelmäßig über betriebswirtschaftliche Tagungen, Seminare und Konferenzen sowie über persönliche Veränderungen vorwiegend an den Hochschulen. Darüber hinaus werden auch Nachrichten für Studenten und Wirtschaftspraktiker veröffentlicht, die Bezug zur Hochschule haben.

Umfassendes, modernes, gut verständliches Standardlehrbuch zur Allgemeinen Betriebswirtschaftslehre

↗

WWW.GABLER.DE

Jean-Paul Thommen / Ann-Kristin Achleitner

Allgemeine Betriebswirtschaftslehre

Umfassende Einführung aus managementorientierter Sicht

6., überarb. u. erw. Aufl. 2009. 1183 S. Geb. EUR 52,90

ISBN 978-3-8349-1325-8

Dieses bewährte Lehrbuch gibt eine umfassende und sehr gut verständliche Einführung in alle unternehmerischen Funktionen aus managementorientierter Sicht.

Für die 6. Auflage wurde die „Allgemeine Betriebswirtschaftslehre" von Thommen/Achleitner erneut umfassend bearbeitet. Sämtliche Teile des Buches wurden aktualisiert und ergänzt. Teil 5: Rechnungswesen wurde grundlegend überarbeitet. Insbesondere wurde der wachsenden Bedeutung der International Financial Reporting Standards (IFRS) Rechnung getragen. Diese werden neu in einem eigenen Kapitel behandelt.

Die klare Strukturierung des Inhalts, die vielen anschaulichen Grafiken sowie das umfangreiche Stichwortverzeichnis ermöglichen auch eine Nutzung des Lehrbuches als Nachschlagewerk.

Jean-Paul Thommen / Ann-Kristin Achleitner

Allgemeine Betriebswirtschaftslehre Arbeitsbuch

Repetitionsfragen - Aufgaben - Lösungen

6., vollst. überarb. Aufl. 2009. 578 S. Br. EUR 39,90

ISBN 978-3-8349-1358-6

Das Arbeitsbuch ergänzt das Lehrbuch „Allgemeine Betriebswirtschaftslehre" von Thommen/Achleitner. Es dient der Wiederholung und Umsetzung von Instrumenten zur Gestaltung unternehmerischer Funktionen. Betriebswirtschaftliche Strukturen und Prozesse werden auf Basis moderner, praxisorientierter Fragestellungen und Lösungen erarbeitet. Das Arbeitsbuch folgt dabei der bewährten Struktur des Lehrbuches. Sämtliche Teile wurden aktualisiert und ergänzt. Teil 5: Rechnungswesen wurde grundlegend überarbeitet. Insbesondere wurde der wachsenden Bedeutung der International Financial Reporting Standards (IFRS) Rechnung getragen.

Einfach bestellen: kerstin.kuchta@gwv-fachverlage.de Telefon +49(0)611. 7878-626

KOMPETENZ IN SACHEN WIRTSCHAFT

GABLER

Änderungen vorbehalten. Erhältlich im Buchhandel oder beim Verlag.

IMPRESSUM/HINWEISE FÜR AUTOREN

Verlag

Gabler/GWV Fachverlage GmbH,
Abraham-Lincoln-Straße 46, 65189 Wiesbaden,
http://www.gabler.de
http://www.zfb-online.de
Geschäftsführer: Dr. Ralf Birkelbach, Albrecht F. Schirmacher
Verlagsbereichsleitung Buch, Journals, Seminare, Online:
Maria Akhavan-Hezavei
Gesamtleitung Produktion: Christian Staral
Gesamtleitung Vertrieb: Gabriel Göttlinger

Editor-in-Chief:
Professor Dr. Dr. h.c. Günter Fandel
FernUniversität in Hagen
Fakultät für Wirtschaftswissenschaft
58084 Hagen
Tel: 02331/987-2625, Fax: 02331/987-2575
E-Mail: ZfB@FernUni-Hagen.de

Administration Manuscript Central™
Sebastian Bartussek, Tel.: 02331/987-2652,
Fax: 02331/987-2575, E-Mail: Sebastian.Bartussek@FernUni-Hagen.de

Redaktion: Annelie Meisenheimer, Tel.: 0611/7878-232,
Fax: 0611/7878-411, E-Mail: Annelie.Meisenheimer@gabler.de

Abonnentenbetreuung: Stefanie Druffelsmeyer, Tel.: 05241/801968,
Fax: 05241/809620

Produktmanagement: Melanie Engelhard-Gökalp,
Tel.: 0611/7878-315, Fax: 0611/7878-440,
E-Mail: melanie.engelhard-goekalp@gabler.de

Anzeigenleitung: Stefan Strussione, Tel.: 0611/7878-157,
Fax: 0611/7878-430, E-Mail: Stefan.Strussione@gwv-fachverlage.de

Anzeigendisposition: Monika Dannenberger,
Tel.: 0611/7878-148, Fax: 0611/7878-430,
E-Mail: Monika.Dannenberger@gwv-fachverlage.de

Es gilt die Anzeigenpreisliste vom 1. 1. 2006.

Produktion / Layout: Sandra Kraut

Bezugsmöglichkeiten: Die Zeitschrift erscheint monatlich. Das Abonnement kann jederzeit zur nächsten erreichbaren Ausgabe schriftlich mit Nennung der Kundennummer gekündigt werden. Eine schriftliche Bestätigung erfolgt nicht. Zuviel gezahlte Beträge für nicht gelieferte Ausgaben werden zurückerstattet. Jährlich können 1 bis 6 Special Issues hinzukommen. Jedes Special Issue wird den Abonnenten mit einem Nachlass von 25% des jeweiligen Ladenpreises gegen Rechnung geliefert.

	Preise Inland:	Preise Ausland:
Einzelheft:	42,– Euro	48,– Euro
Studenten-* / Emeritus-Abo:	76,– Euro	97,– Euro
ausgewählte Verbände:**	185,– Euro	203,– Euro
Privat-Abo:	218,– Euro	247,– Euro
Lehrstuhl-Abo:	247,– Euro	269,– Euro
Bibliotheks-/Unternehmensabo:	427,– Euro	439,– Euro

* Studienbescheinigung
** auf Anfrage beim Verlag

Satzherstellung: Fotosatz-Service Köhler GmbH – Reinhold Schöberl, Würzburg.
Druck und Verarbeitung: Druckerei Krips, Meppel, Niederlande.

Dieser Ausgabe liegt eine Beilage des Verlages C.H. Beck oHG, München, bei. Wir bitten unsere Leserinnen und Leser um Beachtung.

Gedruckt auf säurefreiem und chlorfrei gebleichtem Papier.

Printed in Europe

Hinweise für Autoren

1. Bitte beachten Sie die „Grundsätze und Ziele" der ZfB.

2. Einreichungen werden bei der ZfB ausschließlich über ein Online-Verfahren abgewickelt. Manuskripte – in deutscher oder englischer Sprache – können vom Autor unter http://mc.manuscriptcentral.com/zfb direkt in das Manuskriptverwaltungssystem hochgeladen werden. Hierbei ist insbesondere auf die Wahrung der Anonymität der zur Begutachtung eingereichten Vorlagen zu achten. Der Autor verpflichtet sich mit der Einsendung des Manuskripts unwiderruflich, das Manuskript bis zur Entscheidung über die Annahme nicht anderweitig zu veröffentlichen oder zur Veröffentlichung anzubieten. Diese Verpflichtung erlischt nicht durch Korrekturvorschläge im Begutachtungsverfahren.

3. Um die eingereichten Manuskripte in den Begutachtungsprozess geben bzw. diese im Manuskriptlauf zügig behandeln zu können, wird um Beachtung der folgenden Punkte gebeten: Gesamtlänge des Manuskriptes darf 25 DinA4 nicht überschreiten (bei ca. 3800 Zeichen pro Seite), Schriftart „Times New Roman", Schriftgröße 12, einfacher Zeilenabstand, jeweils 2,5 cm Außenrand, Angabe von Abbildungs- und Tabellenüberschriften (Abb. 1: Text; Tab. 1: Text etc.), eingebundene Objekte (insbes. Bild-, .ppt-, .xls-Dateien etc.) auch separat in Dateiform beifügen, das Hauptdokument muss in **anonymer** Form eingereicht werden, d.h. alle Autorennamen, Autoreninformationen und evtl. Danksagungen sind für die Begutachtung restlos zu streichen. Einhaltung der Gliederungssystematik: **1 Überschriftsebene 1** (12pt, fett, 2 Zeilen Abstand davor, 1 Zeile danach), *1.1 Überschriftsebene 2* (12pt, 1 Zeile Abstand davor, 1 Zeile danach), 1.1.1 Überschriftsebene 3 (12pt, kursiv, 1 Zeile Abstand davor, 1 Zeile danach), **Spitzmarke:** (12pt, fett mit Doppelpunkt zu Beginn des Absatzes, 1 Zeile Abstand davor). Harvard-Zitierweise, keine End- oder Fußnoten: Ein Autor: (vgl. Meier 2007) bzw. (Meier 2007, S. 30); Zwei Autoren: (vgl. Meier/Müller 2007) bzw. (Meier/Müller 2007, S. 30); Drei oder mehr Autoren: (vgl. Meier et al. 2007) bzw. (Meier et al. 2007, S. 30); Eventuelle Erläuterungen zu Textpassagen können weiterhin als Endnoten angehängt werden, sollten aber – soweit möglich – vermieden werden. Das Literaturverzeichnis muss in *Harvard Stil* bzw. *Basic Springer Reference Style* aufgebaut sein. Bei einer Wiedereinreichung eines Beitrags muss eine Stellungnahme zu den Gutachten beigelegt werden. Einreichung der Beitragsdatei als **Microsoft Word®-Datei** oder in einem Word®-kompatiblen Format; **kein (La)TeX. PDF-Dateien sind generell nicht geeignet und können auch nicht ins Onlinesystem Manuscript CentralTM hochgeladen werden**. Der Beitrag muss in folgender Reihenfolge aufgebaut sein: Erste Seite: prägnanter Beitragstitel in deutscher bzw. in englischer Sprache (max. 80 Zeichen; bei Bedarf: Angabe eines Untertitels), dem Beitrag vorgestellte einleitende „Zusammenfassung" bzw. einleitender „Abstract" (Fließtext, max. 15 Zeilen bzw. 1100 Zeichen), deutsche „Schlüsselwörter" (max. 5 Angaben) bzw. englische Keywords (max. 5 Angaben), JEL-Klassifikation (max. 3 Angaben); Ab Seite 2: Beitragstext, falls nötig: „Anmerkungen" als Endnoten (keine Fußnoten im Text), „Literaturverzeichnis", letzte Seite: (nur bei deutschprachigen, enfällt bei englischsprachigen Beiträgen) prägnanter Beitragstitel in englischer Sprache (max. 80 Zeichen, bei Bedarf: Angabe eines Untertitels), „Abstract" in englischer Sprache (Fließtext, max. 15 Zeilen bzw. 1100 Zeichen). Zusätzlich sollten sowohl die Autorenfotos (in digitaler Form, 300dpi, mind. 640x480 Pixel) als auch die Autorenangaben (Titel, Name, Institut, Lehrstuhl, Adresse, Land, ggf. Arbeitsgebiete, Emailadresse und URL; insgesamt pro Autor max. 4 Zeilen) in separaten Dateien eingereicht werden. **Alle Kopf- und Fußzeilen sowie Seitenzahlen sind zu entfernen!**

4. Der Autor verpflichtet sich, die Korrekturfahnen innerhalb einer Woche zu lesen und die Mehrkosten für Korrekturen, die nicht vom Verlag zu vertreten sind, sowie die Kosten für die Korrektur durch einen Korrektor bei nicht termingerechter Rücksendung der Fahnenkorrektur zu übernehmen.

5. Der Autor ist damit einverstanden, dass sein Beitrag außer in der Zeitschrift auch durch Lizenzvergabe in anderen Zeitschriften (auch übersetzt), durch Nachdruck in Sammelbänden (z.B. zu Jubiläen der Zeitschrift oder des Verlages oder in Themenbänden), durch längere Auszüge in Büchern des Verlages auch zu Werbezwecken, durch Vervielfältigung und Verbreitung auf CD-ROM oder anderen Datenträgern, durch Speicherung auf Datenbanken, deren Weitergabe und dem Abruf von solchen Datenbanken während der Dauer des Urheberrechtsschutzes an dem Beitrag im In- und Ausland vom Verlag und seinen Lizenznehmern genutzt wird.

Wege der Personalführung und Unternehmenskultur zur Lösung demographischer Herausforderungen

↗

WWW.GABLER.DE

Heike Bruch / Florian Kunze / Stephan Böhm

Generationen erfolgreich führen

Konzepte und Praxiserfahrungen zum Management des demographischen Wandels

2010. 275 S. Geb. EUR 39,90

ISBN 978-3-8349-1042-4

Die Autoren beschreiben die Auswirkungen des demographischen Wandels auf Unternehmen und geben Handlungsempfehlungen für Personalmanager und Führungskräfte. Dabei werden die Konsequenzen dieses Wandels in Bezug auf verschiedene Ebenen des Unternehmens – individuelle, Team- und Organisationsebene – thematisiert. Alterungsprozesse sowie insbesondere der zunehmende Generationenmix und seine Folgen werden anschaulich beschrieben und interpretiert, verschiedene Lösungsansätze werden vorgestellt. Best-Practice-Beispiele aus deutschen und schweizerischen Unternehmen zeigen praktische Umsetzungsformen und stellen konkrete Umsetzungsmöglichkeiten aus verschiedenen Branchen vor.

Der Inhalt

- Trends der demographischen Entwicklung
- Die Veränderung der Leistungsdispositionen im Alter und individuelle generationale Unterschiede – eine individuelle Perspektive
- Auswirkungen von Altersdifferenzen in der Mitarbeiter-Führungskraft-Beziehung – eine Führungsperspektive
- Das produktive Management von alters- und generationengemischten Teams – eine Teamperspektive
- Der demographische Wandel als strategische Herausforderung im Personalmanagement – eine Gesamtunternehmensperspektive

Die Autoren

Prof. Dr. Heike Bruch ist Professorin und Direktorin am Institut für Führung und Personalmanagement (I.FPM) der Universität St. Gallen. Sie ist Mitglied im wissenschaftlichen Beirat des Demographienetzwerkes (ddn), im Vorstand der DGFP sowie im McKinsey Academic Sounding Board und hat die wissenschaftliche Leitung von TOP JOB inne. Florian Kunze ist wissenschaftlicher Mitarbeiter am I.FPM und Promotionsstudent an der Universität St. Gallen. Dr. Stephan Böhm ist wissenschaftlicher Assistent und Projektleiter am I.FPM. Als Visiting Scholar ist er am Oxford Institute of Ageing der University of Oxford tätig.

Einfach bestellen: kerstin.kuchta@gwv-fachverlage.de Telefon +49(0)611. 7878-626

KOMPETENZ IN SACHEN WIRTSCHAFT

GABLER

Änderungen vorbehalten. Erhältlich im Buchhandel oder beim Verlag.